Mathematics
Teacher's Guide Part 2

CONTENTS

Author: **Carol Bauler, B.A.**

Editor: Alan Christopherson, M.S.

Graphic Design: JoAnn Cumming, A.A.

Alpha Omega Publications®

804 N. 2nd Ave. E., Rock Rapids, IA 51246-1759

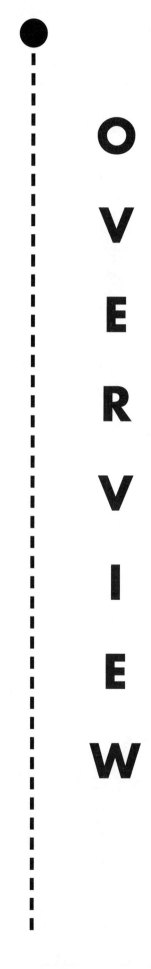

OVERVIEW

MATHEMATICS

Curriculum Overview
Grades K-12

Kindergarten

Lessons

1-40	41-80	81-120	121-160
Directions-right, left, high,low,etc.	**Directions**-right,left, high,low,etc.	**Directions**-right,left, high,low,etc.	**Directions**-right,left, high,low,etc.
Comparisons-big, little,alike,different	**Comparisons**-big, little,alike,different	**Comparisons**-big, little,alike,different	**Comparisons**-big, little,alike,different
Matching	**Matching**	**Matching**	**Matching**
Cardinal Numbers-to 9	**Cardinal Numbers**-to 12	**Cardinal Numbers**-to 19	**Cardinal Numbers**-to 100
Colors-red,blue,green, yellow, brown,purple	**Colors**-orange	**Colors**-black,white	**Colors**-pink
Shapes-circle,square, rectangle,triangle	**Shapes**-circle,square, rectangle,triangle	**Shapes**-circle square, rectangle,triangle	**Shapes**-circle,square, rectangle,triangle
Number Order	**Number Order**	**Number Order**	**Number Order**
Before and After	**Before and After**	**Before and After**	**Before and After**
Ordinal Numbers-to 9th	**Ordinal Numbers**-to 9th	**Ordinal Numbers**-to 9th	**Ordinal Numbers**-to 9th
Problem Solving	**Problem Solving**	**Problem Solving**	**Problem Solving**
	Number Words-to nine	**Number Words**-to nine	**Number Words**-to nine
	Addition-to 9	**Addition**-to 10 and multiples of 10	**Addition**-to 10 and multiples of 10
		Subtraction-to 9	**Subtraction**-to 10
		Place Value	**Place Value**
		Time/Calendar	**Time/Calendar**
			Money
			Skip Counting-2's, 5's, 10's
			Greater/ Less than

	Grade 1	Grade 2	Grade 3
LIFEPAC 1	**NUMBER ORDER, ADD/SUBTRACT** • Number order, skip-count • Add, subtract to 9 • Story problems • Measurements • Shapes	**NUMBERS AND WORDS TO 100** • Numbers and words to 100 • Operation symbols: +, −, =, >, < • Add and subtract • Place value and fact families • Story problems	**ADD/SUB TO 18 AND PLACE VALUE** • Digits, place value to 999 • Add, subtract • Linear measurements • Operation symbols: +, −, =, ≠, >, < • Time
LIFEPAC 2	**ADD/SUBTRACT TO 10, SHAPES** • Add, subtract to 10 • Number words • Place value • Patterns, sequencing, estimation • Shapes	**ADD/SUBTRACT AND EVEN/ODD** • Numbers and words to 200 • Add, subtract, even and odd • Skip-count 2s, 5s, 10s, • Ordinal numbers, fractions, and money • Shapes	**CARRYING AND BORROWING** • Fact families, patterns, and fractions • Add, subtract – carry, borrow • Skip count 2s, 5s, 10s • Money, shapes, lines • Even and odd
LIFEPAC 3	**FRACTIONS, TIME, AND SYMBOLS** • Number sentences • Fractions • Story problems • Time and symbols: =, ≠ • Oral directions	**ADD W/ CARRYING TO 10'S PLACE** • Add w/ carrying to 10's place • Subtract, • Flat shapes, money, A.M./P.M. • Rounding to 10's place • Standard measurements	**FACTS OF ADD/SUB AND FRACTIONS** • Add 3 numbers w/ carrying • Coins, weight, volume, A.M./P.M. • Fractions, • Skip count 3s, subtract w/ borrowing • Oral instructions
LIFEPAC 4	**ADD TO 18, MONEY, MEASUREMENT** • Add to 18 • Skip-count, even and odd • Money • Shapes and measurement • Place value	**NUMBERS/WORDS TO 999, GRAPHS** • Numbers and words to 999 • Add, subtract, and place value • Calendar • Measurements and solid shapes • Making change	**ROUND, ESTIMATE, STORY PROBLEMS** • Place value to 9,999 • Rounding to 10's and estimation • Add and subtract fractions • Roman numerals • 1/4 inch
LIFEPAC 5	**COLUMN ADDITION AND ESTIMATION** • Add 3 numbers – 1-digit • Ordinal numbers • Time and number lines • Estimation and charts • Fractions	**ADD/SUBTRACT TO 100'S PLACE** • Data and bar graphs, shapes • Add, subtract to 100's • Skip-count 3s, place value to 100's • Add fractions • Temperature	**PLANE SHAPES AND SYMMETRY** • Number sentences • Rounding to 100's and estimation • Perimeter and square inch • Bar graph, symmetry, even/odd rules • Temperature
LIFEPAC 6	**NUMBER WORDS TO 99** • Number words to 99 • Add 2 numbers – 2-digit • Symbols: >, < • Fractions, • Shapes	**SUBTRACT W/ BORROWING FROM 10'S** • Measurements • Time and money • Subtract w/ borrowing from 10's place • Add, subtract fractions • Perimeter	**MULTIPLICATION, LINES, AND ANGLES** • Add, subtract to 9,999 • Multiples, times facts for 2 • Area and equivalent fractions • Line graph, segments and angles • Money
LIFEPAC 7	**COUNT TO 200, SUBTRACT TO 12** • Number order and place value • Subtract to 12 • Operation signs • Estimation and time • Graphs	**ADD W/ CARRYING TO 100'S PLACE** • Add w/ carrying to 100's place • Fractions as words • Number order in books • Rounding and estimation	**ADD/SUB MIXED NUMBERS, PROBABILITY** • Times facts for 5, missing numbers • Mixed numbers – add, subtract • Subtract with 0s in minuend • Circle graph, probability • Probability
LIFEPAC 8	**ADD/SUBTRACT TO 18** • Addition, subtract to 18 • Group counting • Fractions, • Time and measurements • Shapes	**VOLUME AND COIN CONVERSION** • Add, subtract, measurements • Group count and "think" answers • Convert coins • Directions – N, S, E, W • Length and width	**MEASUREMENTS AND MULTIPLICATION** • Times facts for 3, 10 – multiples of 4 • Convert units of measurement • Decimals and directions, • Picture graph, missing addend • Length and width
LIFEPAC 9	**SENSIBLE ANSWERS** • Add 3 numbers – 2-digit • Fact families • Sensible answers • Subtract 2 numbers – 2-digit	**AREA/SQUARE MEASUREMENT** • Area and square measurement • Add 3 numbers – 20 digit w/ carrying • Add coins and convert to cents • Fractions, quarter-inch	**MULT, METRICS, AND PERIMETER** • Add, subtract whole numbers, fractions, mixed numbers • Standard measurements, metrics • Operation symbols • Times facts for 4
LIFEPAC 10	**REVIEW** • Add, subtract, place value • Directions – N, S, E, W • Fractions • Patterns	**REVIEW** • Rules for even and odd • Round numbers to 100's place • Time – digital, sensible answers • Add 3 numbers – 3-digit	**PROBABILITY, UNITS, AND SHAPES** • Add, subtract • Rounding to 1,000's, estimation • Probability, equations, and parentheses • Perimeter and area • Times facts 2,3,4,5,10

Grade 4	Grade 5	Grade 6	
WHOLE NUMBERS AND FRACTIONS • Naming whole numbers • Naming fractions • Sequencing patterns • Numbers to 1,000	OPERATIONS OF WHOLE NUMBERS • Operations and symbols • Fraction language • Grouping, patterns, and sequencing • Rounding and estimation	FRACTIONS AND DECIMALS • Number to billions' place • Add and subtract fractions • Add and subtract decimals • Read and write fractions	LIFEPAC 1
WHOLE NUMBERS MULTIPLICATION • Operation symbols • Multiplication – 1-digit multiplier • Fractions – addition and subtraction • Numbers to 10,000	MULTIPLICATION and DIVISION • Multiplication and division • Fractions: +, –, simplify • Plane and solid shapes • Symbol language	FINDING COMMON DENOMINATORS • Prime factors • Fractions with unlike denominators • Exponential notation • Add and subtract mixed numbers	LIFEPAC 2
SEQUENCING AND ROUNDING • Multiplication with carrying • Rounding and estimation • Sequencing fractions • Numbers to 100,000	PERIMETER, AREA, AND DIVISION • Short division • Lowest common multiple • Perimeter and area • Properties of addition	MULTIPLYING MIXED NUMBERS • Multiply mixed numbers • Divide decimals • Bar and line graphs • Converting fractions and decimals	LIFEPAC 3
LINES AND SHAPES • Plane and solid shapes • Lines and line segments • Addition and subtraction • Multiplication with carrying	SHAPES AND DECIMALS • Lines – shapes – circles • Symmetric – congruent – similar • Decimal place value • Properties of multiplication	DIVIDING MIXED NUMBERS • Divide mixed numbers • Area and perimeter • Standard measurements	LIFEPAC 4
DIVISION AND MEASUREMENTS • Division – 1-digit divisor • Families of facts • Standard measurements • Number grouping	MEASUREMENT AND FRACTIONS • Multiply and divide by 10, 100, 1,000 • Standard measurements • Rate problems • Whole number and fraction operations	METRIC MEASURE • Metric measures • Plane and solid shapes • Multi-operation problems • Roman numerals	LIFEPAC 5
DIVISION, FACTORS, AND FRACTIONS • Division – 1-digit with remainder • Factors and multiples • Fractions – improper and mixed • Equivalent fractions	FRACTIONS AND DECIMALS • Multiplication of fractions • Reading decimal numbers • Adding and subtracting decimals • Multiplication – decimals	LCM AND GCF • LCM, GCF • Fraction and decimal equivalents • Percent • Variables, functions, and formulas	LIFEPAC 6
WHOLE NUMBERS AND FRACTIONS • Multiplication – 2-digit multiplier • Simplifying fractions • Averages • Decimals in money problems	DIVISION AND METRIC UNITS • Division – 2-digit divisor • Metric units • Multiplication – mixed numbers • Multiplication – decimals	INTEGERS, RATIO AND PROPORTION • Positive and negative integers • Ratio and proportion • Fractions, decimals and percents • Statistics	LIFEPAC 7
WHOLE NUMBERS AND FRACTIONS • Division, 1-digit divisor • Fractions – unlike denominators • Metric units • Whole numbers: +, –, x, ÷	CALCULATORS AND PRIME FACTORS • Calculators and whole numbers • Calculators and decimals • Estimation • Prime factors	PROBABILITY AND GRAPHING • Probability • Graphs • Metric and standard units • Square root	LIFEPAC 8
DECIMALS AND FRACTIONS • Reading and writing decimals • Mixed numbers – +, – • Cross multiplication • Estimation	FRACTIONS AND DECIMALS • Division – fractions • Division – decimals • Ratios and ordered pairs • Converting fractions to decimals	CALCULATORS AND ESTIMATION • Calculators • Estimation • Geometric symbols and shapes • Missing number problems	LIFEPAC 9
ESTIMATION, CHARTS, AND GRAPHS • Estimation and data gathering • Charts and Graphs • Review numbers to 100,000 • Whole numbers: +, –, x, ÷	PROBABILITY AND CHARTS • Probability and data gathering • Charts and graphs • Review numbers to 100 million • Fractions and decimals: +, –, x, ÷	INTEGERS AND OPERATIONS • Mental arithmetic • Fraction operations • Variables and properties • Number lines	LIFEPAC 10

Mathematics LIFEPAC Overview

	Grade 7	Grade 8	Grade 9
LIFEPAC 1	INTEGERS • Adding and Subtracting Integers • Multiplying and Dividing Integers • The Real Number System	THE REAL NUMBER SYSTEM • Relationships • Other Forms • Simplifying	VARIABLES AND NUMBERS • Variables • Distributive Property • Definition of signed numbers • Signed number operations
LIFEPAC 2	FRACTIONS • Working with Fractions • Adding and Subtracting Fractions • Multiplying and Dividing Fractions	MODELING PROBLEMS IN INTEGERS • Equations with Real Numbers • Functions • Integers • Modeling with Integers	SOLVING EQUATIONS • Sentences and formulas • Properties • Solving equations • Solving inequalities
LIFEPAC 3	DECIMALS • Decimals and Their Operations • Applying Decimals • Scientific Notation • The Metric System	MODELING PROBLEMS WITH RATIONAL NUMBERS • Number Theory • Solving Problems with Rational Numbers • Solving Equations and Inequalities	PROBLEM ANALYSIS AND SOLUTION • Words and symbols • Simple verbal problems • Medium verbal problems • Challenging verbal problems
LIFEPAC 4	PATTERNS AND EQUATIONS • Variable Expressions • Patterns and Functions • Solving Equations • Equations and Inequalities	PROPORTIONAL REASONING • Proportions • Percents • Measurement/Similar Figures	POLYNOMIALS • Addition of polynomials • Subtraction of polynomials • Multiplication of polynomials • Division of polynomials
LIFEPAC 5	RATIOS AND PROPORTIONS • Ratios, Rates, and Proportions • Using Proportions • Fractions, Decimals, and Percents	MORE WITH FUNCTIONS • Solving Equations • Families of Functions • Patterns	ALGEBRAIC FACTORS • Greatest common factor • Binomial factors • Complete factorization • Word problems
LIFEPAC 6	PROBABILTY AND GRAPHING • Probability • Functions • Graphing Linear Equations • Direct Variation	MEASUREMENT • Angle Measures and Circles • Polygons • Indirect Measure	ALGEBRAIC FRACTIONS • Operations with fractions • Solving equations • Solving inequalities • Solving word problems
LIFEPAC 7	DATA ANALYSIS • Describing Data • Organizing Data • Graphing Data and Making Predictions	PLANE GEOMETRY • Perimeter and Area • Symmetry and Reflections • Other Transformations	RADICAL EXPRESSIONS • Rational and irrational numbers • Operations with radicals • Irrational roots • Radical equations
LIFEPAC 8	GEOMETRY • Basic Geometry • Classifying Polygons • Transformations	MEASURE OF SOLID FIGURES • Surface Area • Solid Figures • Volume • Volume of Composite Figures	GRAPHING • Equations of two variables • Graphing lines • Graphing inequalities • Equations of lines
LIFEPAC 9	MEASUREMENT AND AREA • Perimeter • Area • The Pythagorean Theorem	DATA ANALYSIS • Collecting and Representing Data • Central Tendency and Dispersion • Frequency and Histograms • Box-and-Whisker, and Scatter Plots	SYSTEMS • Graphical solution • Algebraic solutions • Determinants • Word problems
LIFEPAC 10	SURFACE AREA AND VOLUME • Solids • Prisms • Cylinders	PROBABILITY • Outcomes • Permutations and Combinations • Probability and Odds • Independent/Dependent Events	QUADRATIC EQUATIONS AND REVIEW • Solving quadratic equations • Equations and inequalities • Polynomials and factors • Radicals and graphing

Grade 10	Grade 11	Grade 12	
A MATHEMATICAL SYSTEM • Points, lines, and planes • Definition of definitions • Geometric terms • Postulates and theorems	SETS, STRUCTURE, AND FUNCTION • Properties and operations of sets • Axioms and applications • Relations and functions • Algebraic expressions	RELATIONS AND FUNCTIONS • Relations and functions • Rules of correspondence • Notation of functions • Types of functions	LIFEPAC 1
PROOFS • Logic • Reasoning • Two-column proof • Paragraph proof	NUMBERS, SENTENCES AND PROBLEMS • Order and absolute value • Sums and products • Algebraic sentences • Number and motion problems	SPECIAL FUNCTIONS • Linear functions • Second-degree functions • Polynomial functions • Other functions	LIFEPAC 2
ANGLES AND PARALLELS • Definitions and measurement • Relationships and theorems • Properties of parallels • Parallels and polygons	LINEAR EQUATIONS AND INEQUALITIES • Graphs • Equations • Systems of equations • Inequalities	TRIGONOMETRIC FUNCTIONS • Definition • Evaluation of functions • Trigonometric tables • Special angles	LIFEPAC 3
CONGRUENCY • Congruent triangles • Corresponding parts • Inequalities • Quadrilaterals	POLYNOMIALS • Multiplying polynomials • Factoring • Operations with polynomials • Variations	CIRCULAR FUNCTIONS AND GRAPHS • Circular functions and special angles • Graphs of sin and cos • Amplitude and period • Phase shifts	LIFEPAC 4
SIMILAR POLYGONS • Ratios and proportions • Definition of similarity • Similar polygons and triangles • Right triangle geometry	RADICAL EXPRESSIONS • Multiplying and dividing fractions • Adding and subtracting fractions • Equations with fractions • Applications of fractions	IDENTITIES AND FUNCTIONS • Reciprocal relations • Pythagorean relations • Trigonometric identities • Sum and difference formulas	LIFEPAC 5
CIRCLES • Circles and spheres • Tangents, arcs, and chords • Special angles in circles • Special segments in circles	REAL NUMBERS • Rational and irrational numbers • Laws of radicals • Quadratic equations • Quadratic formula	TRIGONOMETRIC FUNCTIONS • Trigonometric functions • Law of cosines • Law of sines • Applied problems	LIFEPAC 6
CONSTRUCTION AND LOCUS • Basic constructions • Triangles and circles • Polygons • Locus meaning and use	QUADRATIC RELATIONS AND SYSTEMS • Distance formulas • Conic sections • Systems of equations • Application of conic sections	TRIGONOMETRIC FUNCTIONS • Inverse functions • Graphing polar coordinates • Converting polar coordinates • Graphing polar equations	LIFEPAC 7
AREA AND VOLUME • Area of polygons • Area of circles • Surface area of solids • Volume of solids	EXPONENTIAL FUNCTIONS • Exponents • Exponential equations • Logarithmic functions • Matrices	QUADRATIC EQUATIONS • Conic sections • Circle and ellipse • Parabola and hyperbola • Transformations	LIFEPAC 8
COORDINATE GEOMETRY • Ordered pairs • Distance • Lines • Coordinate proofs	COUNTING PRINCIPLES • Progressions • Permutations • Combinations • Probability	PROBABILITY • Random experiments and probability • Permutations • Combinations • Applied problems	LIFEPAC 9
REVIEW • Proof and angles • Polygons and circles • Construction and measurement • Coordinate geometry	REVIEW • Integers and open sentences • Graphs and polynomials • Fractions and quadratics • Exponential functions	CALCULUS • Mathematical induction • Functions and limits • Slopes of functions • Review of Mathematics 1200	LIFEPAC 10

MANAGEMENT

STRUCTURE OF THE LIFEPAC CURRICULUM

The LIFEPAC curriculum is conveniently structured to provide one teacher handbook containing teacher support material with answer keys and ten student worktexts for each subject at grade levels two through twelve. The worktext format of the LIFEPACs allows the student to read the textual information and complete workbook activities all in the same booklet. The easy to follow LIFEPAC numbering system lists the grade as the first number(s) and the last two digits as the number of the series. For example, the Language Arts LIFEPAC at the 6th grade level, 5th book in the series would be LAN0605.

Each LIFEPAC is divided into 3 to 5 sections and begins with an introduction or overview of the booklet as well as a series of specific learning objectives to give a purpose to the study of the LIFEPAC. The introduction and objectives are followed by a vocabulary section which may be found at the beginning of each section at the lower levels, at the beginning of the LIFEPAC in the middle grades, or in the glossary at the high school level. Vocabulary words are used to develop word recognition and should not be confused with the spelling words introduced later in the LIFEPAC. The student should learn all vocabulary words before working the LIFEPAC sections to improve comprehension, retention, and reading skills.

Each activity or written assignment has a number for easy identification, such as 1.1. The first number corresponds to the LIFEPAC section and the number to the right of the decimal is the number of the activity.

Teacher checkpoints, which are essential to maintain quality learning, are found at various locations throughout the LIFEPAC. The teacher should check 1) neatness of work and penmanship, 2) quality of understanding (tested with a short oral quiz), 3) thoroughness of answers (complete sentences and paragraphs, correct spelling, etc.), 4) completion of activities (no blank spaces), and 5) accuracy of answers as compared to the answer key (all answers correct).

The self test questions are also number coded for easy reference. For example, 2.015 means that this is the 15th question in the self test of Section II. The first number corresponds to the LIFEPAC section, the zero indicates that it is a self test question, and the number to the right of the zero the question number.

The LIFEPAC test is packaged at the centerfold of each LIFEPAC. It should be removed and put aside before giving the booklet to the student for study.

Answer and test keys have the same numbering system as the LIFEPACs and appear at the back of this handbook. The student may be given access to the answer keys (not the test keys) under teacher supervision so that he can score his own work.

A thorough study of the Curriculum Overview by the teacher before instruction begins is essential to the success of the student. The teacher should become familiar with expected skill mastery and understand how these grade level skills fit into the overall skill development of the curriculum. The teacher should also preview the objectives that appear at the beginning of each LIFEPAC for additional preparation and planning.

TEST SCORING and GRADING

Answer keys and test keys give examples of correct answers. They convey the idea, but the student may use many ways to express a correct answer. The teacher should check for the essence of the answer, not for the exact wording. Many questions are high level and require thinking and creativity on the part of the student. Each answer should be scored based on whether or not the main idea written by the student matches the model example. "Any Order" or "Either Order" in a key indicates that no particular order is necessary to be correct.

Most self tests and LIFEPAC tests at the lower elementary levels are scored at 1 point per question; however, the upper levels may have a point system awarding 2 to 5 points for various questions. Further, the total test points will vary; they may not always equal 100 points. They may be 78, 85, 100, 105, etc.

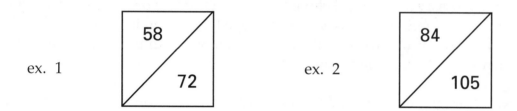

A score box similar to ex.1 above is located at the end of each self test and on the front of the LIFEPAC test. The bottom score, 72, represents the total number of points possible on the test. The upper score, 58, represents the number of points your student will need to receive an 80% or passing grade. If you wish to establish the exact percentage that your student has achieved, find the total points of his correct answers and divide it by the bottom number (in this case 72.) For example, if your student has a point total of 65, divide 65 by 72 for a grade of 90%. Referring to ex. 2, on a test with a total of 105 possible points, the student would have to receive a minimum of 84 correct points for an 80% or passing grade. If your student has received 93 points, simply divide the 93 by 105 for a percentage grade of 89%. Students who receive a score below 80% should review the LIFEPAC and retest using the appropriate Alternate Test found in the Teacher's Guide.

The following is a guideline to assign letter grades for completed LIFEPACs based on a maximum total score of 100 points.

LIFEPAC Test	=	60% of the Total Score (or percent grade)
Self Test	=	25% of the Total Score (average percent of self tests)
Reports	=	10% or 10* points per LIFEPAC
Oral Work	=	5% or 5* points per LIFEPAC

*Determined by the teacher's subjective evaluation of the student's daily work.

Example:

LIFEPAC Test Score	=	92%	92	x	.60	=	55 points
Self Test Average	=	90%	90	x	.25	=	23 points
Reports						=	8 points
Oral Work						=	4 points

TOTAL POINTS = 90 points

Grade Scale based on point system:

100	–	94	=	A
93	–	86	=	B
85	–	77	=	C
76	–	70	=	D
Below		70	=	F

TEACHER HINTS and STUDYING TECHNIQUES

LIFEPAC Activities are written to check the level of understanding of the preceding text. The student may look back to the text as necessary to complete these activities; however, a student should never attempt to do the activities without reading (studying) the text first. Self tests and LIFEPAC tests are never open book tests.

Language arts activities (skill integration) often appear within other subject curriculum. The purpose is to give the student an opportunity to test his skill mastery outside of the context in which it was presented.

Writing complete answers (paragraphs) to some questions is an integral part of the LIFEPAC Curriculum in all subjects. This builds communication and organization skills, increases understanding and retention of ideas, and helps enforce good penmanship. Complete sentences should be encouraged for this type of activity. Obviously, single words or phrases do not meet the intent of the activity, since multiple lines are given for the response.

Review is essential to student success. Time invested in review where review is suggested will be time saved in correcting errors later. Self tests, unlike the section activities, are closed book. This procedure helps to identify weaknesses before they become too great to overcome. Certain objectives from self tests are cumulative and test previous sections; therefore, good preparation for a self test must include all material studied up to that testing point.

The following procedure checklist has been found to be successful in developing good study habits in the LIFEPAC curriculum.

1. Read the introduction and Table of Contents.
2. Read the objectives.
3. Recite and study the entire vocabulary (glossary) list.
4. Study each section as follows:
    a. Read the introduction and study the section objectives.
    b. Read all the text for the entire section, but answer none of the activities.
    c. Return to the beginning of the section and memorize each vocabulary word and definition.
    d. Reread the section, complete the activities, check the answers with the answer key, correct all errors, and have the teacher check.
    e. Read the self test but do not answer the questions.
    f. Go to the beginning of the first section and reread the text and answers to the activities up to the self test you have not yet done.
    g. Answer the questions to the self test without looking back.
    h. Have the self test checked by the teacher.
    i. Correct the self test and have the teacher check the corrections.
    j. Repeat steps a–i for each section.

5. Use the SQ3R* method to prepare for the LIFEPAC test.
6. Take the LIFEPAC test as a closed book test.
7. LIFEPAC tests are administered and scored under direct teacher supervision. Students who receive scores below 80% should review the LIFEPAC using the SQ3R* study method and take the Alternate Test located in the Teacher Handbook. The final test grade may be the grade on the Alternate Test or an average of the grades from the original LIFEPAC test and the Alternate Test.

      *SQ3R: **S**can the whole LIFEPAC.
                  Question yourself on the objectives.
                  Read the whole LIFEPAC again.
                  Recite through an oral examination.
                  Review weak areas.

GOAL SETTING and SCHEDULES

Each school must develop its own schedule, because no single set of procedures will fit every situation. The following is an example of a daily schedule that includes the five LIFEPAC subjects as well as time slotted for special activities.

Possible Daily Schedule

8:15	–	8:25	Pledges, prayer, songs, devotions, etc.
8:25	–	9:10	Bible
9:10	–	9:55	Language Arts
9:55	–	10:15	Recess (juice break)
10:15	–	11:00	Mathematics
11:00	–	11:45	Social Studies
11:45	–	12:30	Lunch, recess, quiet time
12:30	–	1:15	Science
1:15	–		Drill, remedial work, enrichment*

*Enrichment: Computer time, physical education, field trips, fun reading, games and puzzles, family business, hobbies, resource persons, guests, crafts, creative work, electives, music appreciation, projects.

Basically, two factors need to be considered when assigning work to a student in the LIFEPAC curriculum.

The first is time. An average of 45 minutes should be devoted to each subject, each day. Remember, this is only an average. Because of extenuating circumstances a student may spend only 15 minutes on a subject one day and the next day spend 90 minutes on the same subject.

The second factor is the number of pages to be worked in each subject. A single LIFEPAC is designed to take 3 to 4 weeks to complete. Allowing about 3-4 days for LIFEPAC introduction, review, and tests, the student has approximately 15 days to complete the LIFEPAC pages. Simply take the number of pages in the LIFEPAC, divide it by 15 and you will have the number of pages that must be completed on a daily basis to keep the student on schedule. For example, a LIFEPAC containing 45 pages will require 3 completed pages per day. Again, this is only an average. While working a 45 page LIFEPAC, the student may complete only 1 page the first day if the text has a lot of activities or reports, but go on to complete 5 pages the next day.

Long range planning requires some organization. Because the traditional school year originates in the early fall of one year and continues to late spring of the following year, a calendar should be devised that covers this period of time. Approximate beginning and completion dates can be noted on the calendar as well as special occasions such as holidays, vacations and birthdays. Since each LIFEPAC takes 3-4 weeks or eighteen days to complete, it should take about 180 school days to finish a set of ten LIFEPACs. Starting at the beginning school date, mark off eighteen school days on the calendar and that will become the targeted completion date for the first LIFEPAC. Continue marking the calendar until you have established dates for the remaining nine LIFEPACs making adjustments for previously noted holidays and vacations. If all five subjects are being used, the ten established target dates should be the same for the LIFEPACs in each subject.

FORMS

The sample weekly lesson plan and student grading sheet forms are included in this section as teacher support materials and may be duplicated at the convenience of the teacher.

The student grading sheet is provided for those who desire to follow the suggested guidelines for assignment of letter grades found on page 3 of this section. The student's self test scores should be posted as percentage grades. When the LIFEPAC is completed the teacher should average the self test grades, multiply the average by .25 and post the points in the box marked self test points. The LIFEPAC percentage grade should be multiplied by .60 and posted. Next, the teacher should award and post points for written reports and oral work. A report may be any type of written work assigned to the student whether it is a LIFEPAC or additional learning activity. Oral work includes the student's ability to respond orally to questions which may or may not be related to LIFEPAC activities or any type of oral report assigned by the teacher. The points may then be totaled and a final grade entered along with the date that the LIFEPAC was completed.

The Student Record Book which was specifically designed for use with the Alpha Omega curriculum provides space to record weekly progress for one student over a nine week period as well as a place to post self test and LIFEPAC scores. The Student Record Books are available through the current Alpha Omega catalog; however, unlike the enclosed forms these books are not for duplication and should be purchased in sets of four to cover a full academic year.

WEEKLY LESSON PLANNER

Week of:

	Subject	Subject	Subject	Subject
Monday				
	Subject	Subject	Subject	Subject
Tuesday				
	Subject	Subject	Subject	Subject
Wednesday				
	Subject	Subject	Subject	Subject
Thursday				
	Subject	Subject	Subject	Subject
Friday				

WEEKLY LESSON PLANNER

Week of:

	Subject	Subject	Subject	Subject
Monday				
	Subject	Subject	Subject	Subject
Tuesday				
	Subject	Subject	Subject	Subject
Wednesday				
	Subject	Subject	Subject	Subject
Thursday				
	Subject	Subject	Subject	Subject
Friday				

Student Name _____   Year _____

Bible

LP #	Self Test Scores by Sections					Self Test Points	LIFEPAC Test	Oral Points	Report Points	Final Grade	Date
	1	2	3	4	5						
01											
02											
03											
04											
05											
06											
07											
08											
09											
10											

History & Geography

LP #	Self Test Scores by Sections					Self Test Points	LIFEPAC Test	Oral Points	Report Points	Final Grade	Date
	1	2	3	4	5						
01											
02											
03											
04											
05											
06											
07											
08											
09											
10											

Language Arts

LP #	Self Test Scores by Sections					Self Test Points	LIFEPAC Test	Oral Points	Report Points	Final Grade	Date
	1	2	3	4	5						
01											
02											
03											
04											
05											
06											
07											
08											
09											
10											

Student Name _____   Year _____

Mathematics

LP #	Self Test Scores by Sections 1	2	3	4	5	Self Test Points	LIFEPAC Test	Oral Points	Report Points	Final Grade	Date
01											
02											
03											
04											
05											
06											
07											
08											
09											
10											

Science

LP #	Self Test Scores by Sections 1	2	3	4	5	Self Test Points	LIFEPAC Test	Oral Points	Report Points	Final Grade	Date
01											
02											
03											
04											
05											
06											
07											
08											
09											
10											

Spelling/Electives

LP #	Self Test Scores by Sections 1	2	3	4	5	Self Test Points	LIFEPAC Test	Oral Points	Report Points	Final Grade	Date
01											
02											
03											
04											
05											
06											
07											
08											
09											
10											

NOTES

INSTRUCTIONS FOR FIRST GRADE MATHEMATICS

The first grade handbooks of the LIFEPAC curriculum are designed to provide a step-by-step procedure that will help the teacher prepare for and present each lesson effectively. In the early LIFEPACs, the teacher should read the directions and any other sentences to the children. However, as the school year progresses, the student should be encouraged to begin reading and following his own instructional material in preparation for the independent study approach that begins at the second grade level.

This section of the Teacher's Guide includes the following teacher aids:
1) Introduction of Skills   2) Mathematics Terms   3) Teacher Instruction Pages.

The Introduction of Skills is a more detailed overview of skills than that presented in the *Scope and Sequence*. The Mathematics Terms includes a glossary of mathematics terms and a table of measurements. The Teacher Instruction Pages list the Concepts to be taught as well as Student Objectives and Goals for the Teacher. The Teacher Instruction Pages also contain guidelines for teaching each lesson and often include additional learning activities.

Mathematics is a subject that requires skill mastery. But skill mastery needs to be applied toward active student involvement. The Teacher Instruction Pages list the required or suggested materials used in the LIFEPAC lessons. These materials include items generally available in the school or home. Pencils, paper, crayons, scissors, paste and/or glue stick are materials used on a regular basis. Construction paper, beads, buttons, and beans can be used for counting, sets, grouping, fractions, and sequencing. Measurements require measuring cups, rulers, and empty containers. Boxes and similar items help in the study of solid shapes.

Any workbook assignment that can be supported by a real world experience will enhance the student's ability for problem solving. There is an infinite challenge for the teacher to provide a meaningful environment for the study of mathematics. It is a subject that requires constant assessment of student progress. Do not leave the study of mathematics in the classroom.

INTRODUCTION OF SKILLS

Introduction of Skills is a quick reference guide for the teacher who may be looking for a rule or explanation that applies to a particular skill or to find where or when certain skills are introduced in the LIFEPACs. The first number after the skill identifies the LIFEPAC, and the second number identifies the section. 105/3 refers to Mathematics LIFEPAC 105, Section 3.

Addition
facts to 9	101/3
facts to 10	102/1
1-digit number added to 10	102/3
facts to 18	104/1
3 numbers 1-digit	105/1
2 numbers 2-digits	106/3
checking answers	105/1
3 numbers 2-digits	109/1

Calendar   103/4

Count
to 99	101/1
to 100	103/4
to 200	107/1

Directions
    north, east, south, west   110/3

Estimation
size and weight	102/4
numbers	107/5

Even and odd numbers   104/2

Families of facts
    addition and subtraction   109/1

Fractions
1/2 of an object, of a set	103/3
1/4 of an object, of a set	105/3

Graphs (Charts)
    posting data   105/4, 107/5, 109/3

Measurements
objects big and little	101/4
objects greater than and less than	101/4
long and short	101/4
dozen	105/3
ruler - inch	101/4
ruler - one-half inch	108/2
weight	102/4

Money

pennies, dimes	104/4
nickels	106/4

Number line   101/1

Number order

before and after to 99	101/2
bigger and smaller to 99	101/2
before and after to 100	103/5
before and after to 200	107/3
greater than, less than to 100	105/2
greater than, less than to 200	107/1
closest multiple of 10	109/3, 110/1

Number sentences   103/2

Number words

zero to ten	102/5
to twenty	104/3
to ninety-nine	106/2

Operation symbols

+, −, =	102/1
≠	103/2
>, <	106/2

Ordinal numbers

to tenth	102/5

Place value

for ones	102/3
for tens	102/3
for hundreds	107/2

Problem solving

estimation	102/4
how many facts equal a number	102/5
sensible answers	109/4

Sequencing and number patterns   102/5

Shapes

flat	101/4
solid	102/4

Skip counting

by 10's	101/1, 104/2
by 2's	104/2
by 5's	106/1
objects by grouping	108/5

Story problems

oral problems	101/4, 103/3
oral/written	103/5
written	104/4

Subtraction

facts to 9	101/3	
facts to 10	102/2	
facts to 12	107/3	
facts to 18	108/1,3	
1-digit from 10's n/b*	109/4	*n/b no borrowing
2 numbers 2-digits n/b*	109/4	

Time

to hour	103/4
to half-hour	105/3
to quarter-hour	106/4
to 5 minutes	108/2
a.m., p.m.	107/4

Write numbers

to 99	101/1-3
to 200	107/1

Zero as a place holder   107/2

MATHEMATICS TERMS

acute angle   An angle that is less than a right angle or less than 90 degrees.

addend   A number to be added in an addition problem.

angle   The distance between two rays or line segments with a common endpoint.

associative property   No matter how numbers are grouped in addition and multiplication, the answer is always the same.

area   The measurement of a flat surface.   A = l x w (rectangle) A = πr^2 (circle) A = 1/2 b x h (triangle)

average   The total of a group divided by the number in the group.

bar graph   A graph that uses bars to show data.

base   The bottom part of a geometric figure on which the figure rests. The number used as a factor in exponential notation.

cancelling   Simplifying a problem in multiplication or division of fractions within the problem.

cardinal numbers   Numbers used for counting. 1, 2, 3, 4.....

Celsius   Metric unit of measurement for temperature. Freezing-0° C., Boiling 100 °C.

chart   An arrangement of data in a logical order.

circle   A continuous closed line always the same distance from a center point.

circle graph   A circular graph that always represents the whole of the data.

circumference   The distance around (perimeter) a circle.   C = $2\pi r$ or C = πd

common denominator   Fractions must have the same or common denominator to be added or subtracted.

commutative property   No matter what order numbers are added or multiplied, the answer is always the same.

compass   An instrument having two hinged legs used for drawing circles, curved lines, and measuring distances.

composite number   A number that can be divided by 1, by itself, and other numbers.

congruent   Figures that have the same size and shape.

cross multiplication   Multiplying the numerators and denominators of two fractions.

cube   A solid shape with six square faces.

cylinder   A round shape with flat ends.

data   A list of facts from which a conclusion may be drawn.

decimal number   A fraction with an understood denominator of 10, 100, 1,000...

decimal point   A dot separating the whole number from the fractional part of a decimal number.

degree   The unit of measurement for angles.

denominator   The bottom number of a fraction. This number represents the whole.

diameter   The distance across a circle straight through the middle.

difference   The answer to a subtraction problem.

digit   Symbols 0, 1, 2, 3, 4, 5 ,6, 7, 8, 9 which when used alone or in combinations represent a value.

dividend   The number being divided in a division problem.

division bar   The line that separates the numerator from the denominator of a fraction.

divisor   The number doing the dividing in a division problem.

endpoints   Dots that show the beginning and end of a line segment.

equal to   Has the same value as. equal = (not equal ≠)

equation A number sentence that contains an equal sign.
equilateral triangle A triangle whose sides are all equal in length.
equivalent fractions Two or more fractions of equal value. To make an equivalent fraction, multiply or divide the numerator and denominator by the same number.
estimate To find an approximate answer.
even number Any number divisible by two.
expanded form Expressing a number by showing the sum of the digits times the place value of each digit.
exponent The number that tells how many times a base number is used as a factor.
exponential notation Writing a number with a base and its exponent.
face The surfaces of a solid figure.
factor(s) Numbers which when multiplied together form a product or multiple.
Fahrenheit U.S. standard measurement for temperature. Freezing 32°F. Boiling 212°F.
fraction A number that represents all or part of a whole.
fraction bar Also called the division bar.
frequency distribution The number of times data falls within a particular classification.
gram Metric unit of the measurement of weight.
graph A special kind of chart. The most common are bar, line, picture, and circle.
greater than Has larger value than. 2>1
greatest common factor The largest factor that can be divided into two numbers.
hexagon A six-sided polygon.
horizontal Level to or parallel to the horizon.
improper fraction A fraction that is greater than or equal to 1. The numerator is larger than or equal to the denominator.
input Data entered into a calculator (computer).
International Date Line The 180th meridian. People who cross the line going west, gain a day. People who cross going east, lose a day.
intersecting lines Lines that cross each other.
invert To turn around the positions of the numerator and denominator of a fraction.
isosceles triangle A triangle that has two sides of equal length.
least common multiple The smallest multiple that two numbers have in common.
less than Has smaller value than. 1<2
line A continuous set of dots that has no beginning and no end.
line graph A graph that shows data by connecting points with lines.
line segment The part of a line that has a beginning and an end.
liter Metric unit of liquid or dry measurement.
minuend The number from which another number is being subtracted in a subtraction problem.
mean The same as the average.
median The number located exactly in the middle of a list of numbers.
meter Metric unit of linear (line) measurement.

Metric Chart of Prefixes

smallest	_milli_	- a unit contains 1,000
	centi	- a unit contains 100
	deci	- a unit contains 10
	unit	- unit (meter, liter, gram)
	deca	contains 10 units
	hecto	contains 100 units
largest	_kilo_	contains 1,000 units

English System of Weights and Measures

Length	Weight	Dry Measure	Liquid Measure
12 inches = 1 foot	16 ounces = 1 pound	2 cups = 1 pint	16 fl ounces = 1 pint
3 feet = 1 yard	2,000 lb = 1 ton	2 pints = 1 quart	2 cups = 1 pint
36 inches = 1 yard		8 quarts = 1 peck	2 pints = 1 quart
5,280 ft = 1 mile		4 pecks = 1 bushel	4 quarts = 1 gallon
320 rods = 1 mile			

Conversion Chart

To convert	To	Multiply by	To convert	To	Multiply by
linear measure					
centimeters	inches	.394	inches	centimeters	2.54
meters	yards	1.0936	yards	meters	.914
kilometers	miles	.62	miles	kilometers	1.609
liquid measure					
liters	quarts	1.057	quarts	liters	.946
dry measure					
liters	quarts	.908	quarts	liters	1.101
weight					
grams	ounces	.0353	ounces	grams	28.35
kilograms	pounds	2.2046	pounds	kilograms	.4536

mode   The number that appears most often in a list of numbers.

mixed number   A number that combines a whole number and a fraction.

multiple   A multiple of a number is a product of that number.

multiplicand   The number being multiplied in a multiplication problem.

multiplier   The number doing the multiplying in a multiplication problem.

negative number   A number with a value less than zero.

norm   A standard for a particular group.

number line   A line with even spaces used to represent certain values.

numeral   A figure that stands for or represents a number.

numerator   The top number of a fraction. This number represents the parts being described.

obtuse angle   An angle greater than a right angle (90 degrees) but less than a straight line (180 degrees).

octagon   An eight-sided polygon.

odd number   Any number that cannot be divided by two.

ordered pairs   Two numbers written in a particular order so that one can be considered the first number and the other the second number.

ordinal numbers   Numbers that show position. 1st, 2nd, 3rd, 4th.....

output   The answer to data entered into a calculator (computer).

oval   A flattened circle - egg shaped.

parallel lines   Lines that are always the same distance apart.

pattern   A set arrangement or design of forms, colors or numbers.

pentagon   A five-sided polygon.

percent   The relationship between a part and a whole. The whole is always 100.

perimeter   The distance around the outside of a closed figure.

perpendicular lines   Lines that form right or 90 degree angles.

pictograph   A graph that uses pictures to represent data.

pi (π)   3.14   Used to solve for the circumference or area of a circle.

place value   The value of a digit determined by its position in a number.

plane shape   A flat shape. A plane shape is two-dimensional.

point of intersection   The one and only point that intersecting lines have in common.

polygon   A closed plane figure with three or more sides.

positive number   A number with a value greater than zero.

prediction   To tell something in advance.

prime factorization   Prime factors of a number expressed in exponential notation.

prime meridian   The longitudinal meridian (0 degrees) that passes through Greenwich, England.

prime number   A number divisible by only 1 and itself.

probability   The study of the likelihood of events.

product   The answer to a multiplication problem.

proper fraction   A fraction greater than 0 but less than 1. The numerator is smaller than the denominator.

property of zero   In addition, any number added to zero will have itself as an answer. In multiplication, any number multiplied by zero will have zero as an answer.

proportion   An equation stating that two ratios are equal.

protractor   A semi-circular instrument marked in degrees used to find the measure of an angle.

pyramid   A solid figure with a polygon as a base and triangular faces that meet at a point.

quadrilateral   A four-sided polygon.

quotient   The answer to a division problem.

radius   The distance from the center of a circle to the edge of a circle. The radius is half of the diameter.

random sample   A sample in which every member of a large group has an equal chance of being chosen.

ratio   The relationship of two numbers to each other written 1:2 or 1/2 .

ray   A line with one endpoint.

reciprocal   The fraction that results from inverting a fraction.
rectangle   A four-sided polygon with four right angles.
rectangular solid   A solid figure with six rectangular faces.
reduced fraction   A fraction equivalent to another fraction that has been written in smaller numbers. This is also called simplifying a fraction or reducing to lowest terms.
remainder   The amount that remains when a division problem has been completed.
right angle   An angle that measures 90 degrees.
right triangle   A triangle with one right angle.

Roman numerals   The ancient Roman numeral system.
I = 1   V = 5   X = 10   L = 50   C = 100   D = 400   M = 1,000

scalene triangle   A triangle with no equal sides.
sequence   Numbers arranged in a certain pattern.
similar   Figures that have the same shape but not necessarily the same size.
solid shape   A shape that takes up space. A solid shape is three-dimensional.
sphere   A geometric solid in a round shape.
square   A rectangle with all sides equal.
straight angle   An angle that measures 180 degrees.
subtrahend   The number being taken away or subtracted in a subtraction problem.
symmetry   Shapes with equal halves.
sum   The answer to an addition problem.
triangle   A three-sided polygon.
vertex   The point at which two rays or line segments meet.
vertical   Straight up and down. Perpendicular to the horizon.
volume   The measurement of space that a solid figure occupies. $V = l \times w \times h$
whole numbers   Digits arranged to represent a value equal to or greater than a whole.

Page 1: Fun with Numbers

CONCEPT(S): purpose of LIFEPAC, objectives

TEACHER GOAL(S): To teach the children
To know what is expected of the student in the LIFEPAC, and
To write first and last names correctly in manuscript.

MATERIALS/MANIPULATIVES:
pencils

TEACHING PAGE 1:

Turn to page 1. Point to the title and the memory verse and read them aloud. Allow time for the children to look through the LIFEPAC. Write the word *OBJECTIVES* on the board and have the children find the word on the page. Explain that the objectives tell the things the students will be expected to do in the LIFEPAC. Read each one and have the children repeat as they run their fingers under the sentence from left to right. Talk about the objectives so that the children will understand what they will be doing. Have each child write his name on the line.

FUN WITH NUMBERS

 My name is   Teacher check

Memory Verse
"...Ask, and it shall be given you; seek, and ye shall find; knock, and it shall be opened unto you."

Matthew 7:7

 **Objectives**

1. I can write numbers as words to ninety-nine.
2. I can understand operation symbols =, ≠, >, <.
3. I can add tens and ones, and tens and tens.
4. I can tell time to the quarter-hour.
5. I can count pennies, nickels, and dimes.
6. I can tell how many facts equal a number.

page 1 (one)

I. PART ONE

Pages 2 and 3: Addition and Subtraction Facts

CONCEPT(S): addition facts to 18, subtraction facts to 10

TEACHER GOAL(S):To teach the children To review addition facts to 18 and subtraction facts to 10.

MATERIALS/MANIPULATIVES:
pencils

TEACHING PAGES 2 and 3:

Turn to page 2. Point to the number line and read the directions with the students. Review addition using the number line. Tell the students that this is a page of addition facts. They should use the number line only when necessary. Discuss a reasonable time for the completion of the page and then let the students work independently.

Turn to page 3. Review subtraction using the number line. Tell the students that this is a page of subtraction facts. They should use the number line only when necessary. Discuss a reasonable time for the completion of the page and then let the students work independently.

These pages should indicate to both the student and the teacher the amount of progress the students are making in committing addition and subtraction facts to memory. Regular drill on facts should continue for all students.

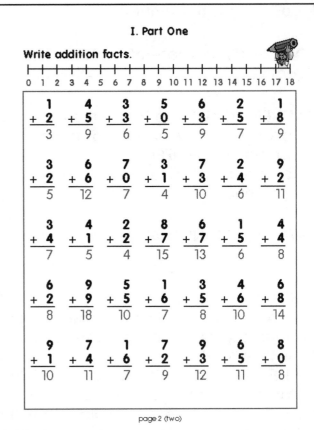

I. Part One

Write addition facts.

1 +2 = 3	4 +5 = 9	3 +3 = 6	5 +0 = 5	6 +3 = 9	2 +5 = 7	1 +8 = 9
3 +2 = 5	6 +6 = 12	7 +0 = 7	3 +1 = 4	7 +3 = 10	2 +4 = 6	9 +2 = 11
3 +4 = 7	4 +1 = 5	2 +2 = 4	8 +7 = 15	6 +7 = 13	1 +5 = 6	4 +4 = 8
6 +2 = 8	9 +9 = 18	5 +5 = 10	1 +6 = 7	3 +5 = 8	4 +6 = 10	6 +8 = 14
9 +1 = 10	7 +4 = 11	1 +6 = 7	7 +2 = 9	9 +3 = 12	6 +5 = 11	8 +0 = 8

page 2 (two)

Write subtraction facts.

5 −5 = 0	7 −2 = 5	4 −1 = 3	5 −3 = 2	5 −4 = 1	4 −2 = 2	1 −0 = 1	4 −4 = 0
3 −1 = 2	4 −0 = 4	6 −3 = 3	8 −2 = 6	3 −3 = 0	4 −2 = 2	6 −1 = 5	9 −1 = 8
8 −1 = 7	2 −1 = 1	6 −4 = 2	5 −1 = 4	3 −2 = 1	3 −1 = 2	4 −3 = 1	7 −3 = 4
10 −2 = 8	3 −0 = 3	8 −7 = 1	8 −2 = 6	6 −1 = 5	10 −4 = 6	7 −2 = 5	8 −7 = 1
7 −4 = 3	6 −3 = 3	2 −1 = 1	8 −4 = 4	9 −5 = 4	8 −0 = 8	5 −3 = 2	4 −2 = 2

page 3 (three)

Pages 4 and 5: Skip Counting, Odd and Even Numbers

CONCEPT(S): counting by 2's, 5's, and 10's, odd and even numbers

TEACHER GOAL(S): To teach the children
    To count by 2's, 5's, and 10's, and
    To recognize odd and even numbers.

MATERIALS/MANIPULATIVES:
pencils, chart of numbers 1 to 99 - LIFEPAC 101 page 7 (or any chart displaying numbers 1-99), addition fact cards through 5's

TEACHING PAGES 4 and 5:
    Review skip counting with the children using their chart of numbers. Tell them to start at different points on the chart and count by 2's, 5's, and 10's.
    Turn to pages 4 and 5. Read each set of directions with the students. Have them complete each exercise independently. Those students who are having difficulty may use their chart of numbers.
    Review *odd* and *even* numbers. Remind the students that they say the *even* numbers when they count by *2's*. The students should begin to recognize that all *even* numbers end in *0, 2, 4, 6, 8* and that all *odd* numbers end in *1, 3, 5, 7, 9*. Use the chart of numbers to point to several different numbers and ask the students to identify the numbers as *odd* or *even*. Complete page 5. Review addition fact cards through *5's*.

Skip Counting

Count by 10!

10, _20_, 30, _40_, _50_,
60, _70_, 80, _90_,

Count by 2!

2,	_4_,	6,	8,	_10_,
12,	14,	_16_,	18,	20,
_22_,	24,	26,	_28_,	30,
32,	_34_,	36,	38,	_40_,
_42_,	44,	_46_,	48,	50,
52,	_54_,	56,	58,	60,
62,	64,	_66_,	68,	_70_,
_72_,	74,	76,	_78_,	80,
82,	_84_,	86,	_88_,	90,
_92_,	94,	_96_,	98	

page 4 (four)

Count by 5!

5, _10_, 15, 20, _25_, _30_,
35, 40, _45_, 50, 55, _60_,
_65_, 70, _75_, _80_, 85, _90_

Circle the odd numbers.

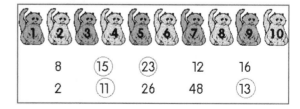

8   (15)   (23)   12   16
2   (11)   26   48   (13)

Circle the even numbers.

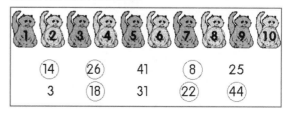

(14)   (26)   41   (8)   25
3   (18)   31   (22)   (44)

page 5 (five)

Page 6: Addition

CONCEPT(S): columnar addition

TEACHER GOAL(S): To teach the children To add numbers in columns.

MATERIALS/MANIPULATIVES:
pencils, addition fact cards for 6's
and 7's

TEACHING PAGE 6:

Turn to page 6. Review addition of three numbers with the students. Introduce the word *sum* to the students. Tell them the *sum* is the name of the answer in addition problems. Remind the students to check their answers by adding down and adding up. Review addition facts for 6's and 7's.

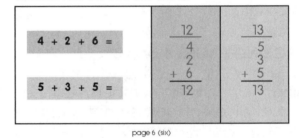

Add down. Check your answer. Add up.

10	9	9	14	13
1	2	5	6	4
5	6	3	3	2
+ 4	+ 1	+ 1	+ 5	+ 7
10	9	9	14	13

10	14	13	15	13
3	6	3	7	9
2	1	4	0	1
+ 5	+ 7	+ 6	+ 8	+ 3
10	14	13	15	13

Rewrite. Find the sum and check.

4 + 2 + 6 =	12	13
	4	5
	2	3
5 + 3 + 5 =	+ 6	+ 5
	12	13

page 6 (six)

SELF TEST 1

CONCEPT(S): addition facts, odd and even numbers, columnar addition

TEACHER GOAL(S): To teach the children To learn to check their progress periodically.

MATERIALS/MANIPULATIVES: pencils, addition fact cards for 8's and 9's

TEACHING PAGE 7:

Turn to page 7. Read the directions with the children. Be sure they understand what they are to do. You may repeat the directions but give no other help. Do not have the children check their own work. Check it as soon as you can and go over it with each child. Show him where he did well and where he needs extra help. Review addition fact cards for *8's* and *9's*.

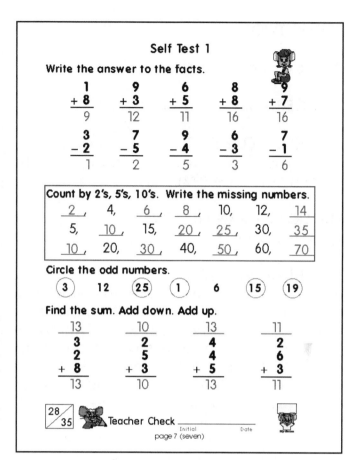

41

II. PART TWO

Pages 8 and 9: Numbers as Words

CONCEPT(S): number words to 99

TEACHER GOAL(S): To teach the children To read and write number words to ninety-nine (99).

MATERIALS/MANIPULATIVES:
pencils, paper

TEACHING PAGES 8 and 9:

Turn to page 8. Have the students read the number words to *nineteen* aloud. Point to the numbers *10* through *90*. Have the children say the numbers aloud. Remind them that this is skip counting by *10's*. Have the students use a piece of paper to cover the number symbols and ask them to say the number words aloud. Point out that the *u* is dropped in the spelling of the word *forty*. Compare the location of the *r* in *three* to its position in the word *thirty*. Ask the students to point to the words *twenty* and *three*. Tell them that we put a hyphen between the two words to join them together and make them one word. Tell the students to write the word *forty* on a piece of paper. Have them write the word *six* next to it. Explain to the students that these numbers mean two different sets - one set of *forty* and one set of *six*. They must add a hyphen between the words to show they mean one set of *forty-six*. Point to the word *seventy-one* on page 8 and have the students say it aloud. Tell them to point to the hyphen that joins the two words together. Have the students close their LIFEPACs. Dictate the words twenty, thirty, forty, fifty, sixty, seventy, eighty, and ninety to them to check how well they do with their spelling.

Turn to page 9. Read the directions and have the students complete the page.

II. Part Two

Write numbers to ninety-nine.

1	— one		11	— eleven
2	— two		12	— twelve
3	— three		13	— thirteen
4	— four		14	— fourteen
5	— five		15	— fifteen
6	— six		16	— sixteen
7	— seven		17	— seventeen
8	— eight		18	— eighteen
9	— nine		19	— nineteen

10	— ten		23	— twenty-three
20	— twenty		46	— forty-six
30	— thirty		71	— seventy-one
40	— forty			
50	— fifty			
60	— sixty			
70	— seventy			
80	— eighty			
90	— ninety			

page 8 (eight)

Write the words on the line. Match.

38	thirty-eight	88	ninety
21	twenty-one	99	seventy-six
47	forty-seven	37	twenty-nine
83	eighty-three	78	forty-two
99	ninety-nine	90	forty-five
25	twenty-five	29	eighty-eight
54	fifty-four	42	thirty-three
66	sixty-six	45	ninety-nine
92	ninety-two	76	thirty-seven
80	eighty	63	forty-nine
75	seventy-five	49	seventy-eight
19	nineteen	33	sixty-three

Write the missing number word.

Three plus eight equals ___eleven___.

Nine plus seven equals ___sixteen___.

___Seven___ plus five equals twelve.

Six plus ___four___ equals ten.

page 9 (nine)

42

Page 10: Numbers as Words

CONCEPT(S): number words to 99

TEACHER GOAL(S): To teach the children
To read and write number words
to ninety-nine (99).

MATERIALS/MANIPULATIVES:
pencils, subtraction fact cards for 0
through 5's

TEACHING PAGE 10:
Turn to page 10. Have the students read the numbers at the top of the page aloud. Read the directions. Explain to the children that this time they should read the number word and then write the number symbol. When the students have completed the page, have them read all the number words aloud. Review the subtraction fact cards through 5's with the students.

20	— twenty	60	— sixty
30	— thirty	70	— seventy
40	— forty	80	— eighty
50	— fifty	90	— ninety

Write the number on the line.

twenty-three	23	sixty-eight	68
forty-five	45	ninety-six	96
fifty-one	51	eighty	80
seventy	70	seventy-five	75
thirty-two	32	fifty	50
seventy-nine	79	thirty-eight	38
sixty	60	sixty-six	66
thirty-four	34	twenty-nine	29
twenty-five	25	forty-seven	47
forty-two	42	sixty-three	63

Write the missing number word.

Ten minus nine equals ___one___.

Eight minus five equals ___three___.

Nine minus ___two___ equals seven.

___Four___ minus four equals zero.

page 10 (ten)

Page 11: Numbers as Words

CONCEPT(S): number sentences

TEACHER GOAL(S): To teach the children To write number sentences using three related numbers.

MATERIALS/MANIPULATIVES:
pencils, paper, subtraction fact cards for 6's and 7's

TEACHING PAGE 11:

Write the numbers *3, 5,* and *8* on the board and ask the students if there is a relationship between these three numbers. Some of the students should be able to tell that *3 + 5 = 8.* Explain to the students that they have made a number sentence out of the three numbers. Have them write the number sentence on a piece of paper in number symbols. Be sure they are using the correct signs. Write the number words *eight, seven,* and *one* on the board. Ask if anyone sees a relationship between these numbers. Again, some students should be able to say, "Eight minus seven equals one." Ask the students to write this number sentence on paper in number words.

Turn to page 11. Point to the top of the page and review what the students have just written on their papers. Read the directions. Remind the children that their number sentences may be addition or subtraction, but they must use the given numbers in the order they are shown. The exercises 1-5 may be written in number symbols. The exercises 6-10 should be written in number words. Be sure the students are using the correct signs and words for signs. Students having difficulty may use their fact cards to find the number combinations. Review subtraction fact cards for *6's* and *7's.*

3, 5, 8	eight, seven, one
3 + 5 = 8	Eight minus seven equals one.

Write a number sentence for each set of numbers.

Number Symbols

1. 8, 4, 4 $8 - 4 = 4$
2. 3, 6, 9 $3 + 6 = 9$
3. 9, 8, 1 $9 - 8 = 1$
4. 6, 7, 13 $6 + 7 = 13$
5. 5, 9, 14 $5 + 9 = 14$

Number Words

6. five, one, four
     Five minus one equals four.

7. six, three, three
     Six minus three equals three.

8. four, eight, twelve
     Four plus eight equals twelve.

9. seven, eight, fifteen
     Seven plus eight equals fifteen

10. ten, six, four
     Ten minus six equals four.

page 11 (eleven)

Page 12: Equal To, Not Equal To, Greater Than, Less Than

CONCEPT(S): equal, not equal, greater than, less than

TEACHER GOAL(S): To teach the children
To write number sentences using the signs equal (=) and not equal (≠), and
To learn the symbols for greater than (>) and less than (<).

MATERIALS/MANIPULATIVES:
pencils

TEACHING PAGE 12:
Turn to page 12. Have the students read the first set of number facts and say them aloud. Ask them if the number sentence 6 + 2 = 4 + 4 is true. Point to the next set of number facts and have the students say them aloud. Ask them if the number sentence 3 + 1 ≠ 2 + 4 is true. Read the first set of directions. Tell the children they must circle the correct symbol to make the sentence true.

Write the number symbols for greater than (>) and less than (<) on the board and explain their meaning. Tell the children that the open side of the symbol should always be toward the larger number; the closed side should always point to the smaller number. Point to the second exercise on page 12. Help the children complete the first column of numbers; then let them complete the second column independently.

Equal (=) and Not Equal (≠)

6 + 2 = 8	4 + 4 = 8

6 + 2 (= , ≠) 4 + 4

3 + 1 = 4	2 + 4 = 6

3 + 1 (= , ≠) 2 + 4

Circle the answer.

5 + 7 (= , ≠) 6 + 6	7 – 3 (= , ≠) 8 – 5
4 + 5 (= , ≠) 7 + 1	10 – 5 (= , ≠) 6 – 1
7 + 3 (= , ≠) 5 + 5	5 – 3 (= , ≠) 6 – 2
8 + 4 (= , ≠) 6 + 3	9 – 3 (= , ≠) 8 – 2
6 + 6 (= , ≠) 4 + 8	10 – 8 (= , ≠) 4 – 0

Greater than (>) Less than (<)

  (> , <)

Circle the answer.

16 (> , <) 15	20 (> , <) 18
17 (> , <) 24	14 (> , <) 18
32 (> , <) 26	27 (> , <) 39
13 (> , <) 18	7 (> , <) 6
49 (> , <) 43	55 (> , <) 50

page 12 (twelve)

Page 13: Number Order

CONCEPT(S): number order to 99

TEACHER GOAL(S): To teach the children
To write numbers in number order to 99,
To write the numbers before and after, and
To write the number between.

MATERIALS/MANIPULATIVES:
pencils, chart of numbers, subtraction facts for 8's and 9's

TEACHING PAGE 13:
Using the chart of numbers, review number order with the students. Remind them that numbers are arranged by the tens' place first and then the ones' place. Name several numbers and ask the students to give the numbers before and after. Say two numbers and ask for the number between. When you are sure the students understand the process have them turn to page 13. Read the instructions aloud with the students and have them complete the page independently. Review the subtraction facts for 8's and 9's.

Write the numbers in number order.

16 19 23 42 45 51 63 76 85 95

Write the number before and the number after.

56	57	58	21	22	23
68	69	70	98	99	100
74	75	76	39	40	41
17	18	19	85	86	87

Write the number between.

12	13	14	18	19	20
26	27	28	77	78	79
59	60	61	80	81	82
43	44	45	9	10	11

page 13 (thirteen)

Page 14: Addition Facts

CONCEPT(S): addition facts to 18

TEACHER GOAL(S): To teach the children
To learn addition facts to 18, and
To recognize colors.

MATERIALS/MANIPULATIVES:
pencils, crayons

TEACHING PAGE 14:
Turn to page 14. Read the rhyme with the students. Have them point out the sailboats and the ducks. Read the directions. When the facts are completed, allow the students to color the boats. Talk to them about their color selection. Be sure they can identify the colors they are using.

A race, a race, a sailboat race,
There is a sailboat race!
But what are those ducks doing,
In the sailboat race?

It may be they are playing,
It may be they don't know,
That ducks should not be swimming,
Where the sailboats go.

Write the answer in the sailboat. Color.

$$\begin{array}{r} 8 \\ +\,2 \\ \hline 10 \end{array}$$

$$\begin{array}{r} 6 \\ +\,6 \\ \hline 12 \end{array}$$

$$\begin{array}{r} 7 \\ +\,4 \\ \hline 11 \end{array}$$

$$\begin{array}{r} 8 \\ +\,6 \\ \hline 14 \end{array}$$

$$\begin{array}{r} 5 \\ +\,5 \\ \hline 10 \end{array}$$

$$\begin{array}{r} 7 \\ +\,5 \\ \hline 12 \end{array}$$

$$\begin{array}{r} 2 \\ +\,9 \\ \hline 11 \end{array}$$

$$\begin{array}{r} 9 \\ +\,7 \\ \hline 16 \end{array}$$

$$\begin{array}{r} 6 \\ +\,4 \\ \hline 10 \end{array}$$

$$\begin{array}{r} 9 \\ +\,5 \\ \hline 14 \end{array}$$

$$\begin{array}{r} 7 \\ +\,8 \\ \hline 15 \end{array}$$

page 14 (fourteen)

SELF TEST 2:

CONCEPT(S): number words, equal, not equal, greater than, less than

TEACHER GOAL(S): To teach the children To learn to check their progress periodically.

MATERIALS/MANIPULATIVES:
pencils

TEACHING PAGE 15:

Turn to page 15. Read the directions with the children. Be sure they understand what they are to do. You may repeat the directions but give no other help. Do not have the children check their own work. Check it as soon as you can and go over it with each child. Show him where he did well and where he needs extra help.

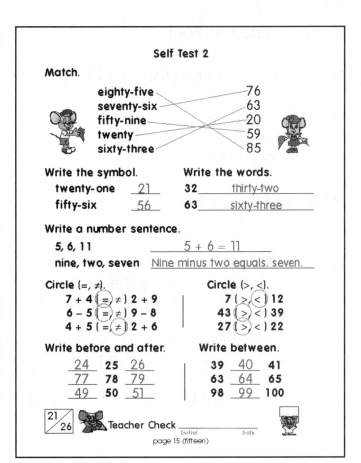

Self Test 2

Match.

eighty-five — 76
seventy-six — 63
fifty-nine — 20
twenty — 59
sixty-three — 85

Write the symbol.   Write the words.
twenty-one   _21_   32   _thirty-two_
fifty-six   _56_   63   _sixty-three_

Write a number sentence.
5, 6, 11   _5 + 6 = 11_
nine, two, seven   _Nine minus two equals. seven._

Circle (=, ≠).   Circle (>, <).
7 + 4 (=) ≠) 2 + 9   7 (>) <) 12
6 − 5 (=) ≠) 9 − 8   43 (>) <) 39
4 + 5 (= (≠) 2 + 6   27 (>) <) 22

Write before and after.   Write between.
_24_ 25 _26_   39 _40_ 41
_77_ 78 _79_   63 _64_ 65
_49_ 50 _51_   98 _99_ 100

21/26 Teacher Check _____
                     Initial   Date
page 15 (fifteen)

III. PART THREE

Page 16: Addition of Tens' Place

CONCEPT(S): addition of tens' place

TEACHER GOAL(S): To teach the children
  To learn to add numbers to tens' place.

MATERIALS/MANIPULATIVES:
pencils, objects for counting (objects that represent ones and objects that represent tens)

TEACHING PAGE 16:

Place the objects that represent *tens* in one group and those that represent *ones* in another group. Give the students *four* objects for *tens* and ask how many? (40) Give the students *five* objects for *ones* and ask how many? (5) Tell the students to add (count) the two sets together and say how many. (45)

Turn to page 16. Have the students read the problem in the illustration aloud. *Forty plus five equals forty-five.* Ask the students to identify the numbers in the ones' place (0, 5) and the tens' place (4). Explain to the children that numbers in problems are always lined up by place value. Tell them we add the ones' place first and the tens' place second. *Be sure the students establish this habit or they will have difficulty in learning carrying in addition.* Complete the first row of problems with the students. They should be able to complete the remainder of the page independently.

III. Part Three

Add tens. $\begin{array}{r} 40 \\ + 5 \\ \hline 45 \end{array}$ Add ones to ones.

4 ⟶   45 ⟵ 0 + 5 = 5

Write the answer.

20 + 2 22	30 + 6 36	40 + 9 49	60 + 3 63	90 + 2 92	10 + 5 15
50 + 3 53	70 + 5 75	80 + 7 87	60 + 4 64	40 + 8 48	30 + 1 31
45 + 3 48	63 + 3 66	72 + 6 78	91 + 3 94	64 + 4 68	22 + 2 24
75 + 2 77	66 + 1 67	45 + 3 48	13 + 2 15	32 + 3 35	88 + 1 89
34 + 3 37	47 + 2 49	52 + 5 57	61 + 5 66	82 + 6 88	96 + 3 99

page 16 (sixteen)

Page 17: Addition of Tens' Place

CONCEPT(S): addition of tens' place

TEACHER GOAL(S): To teach the children
To learn to add numbers to tens' place.

MATERIALS/MANIPULATIVES:
pencils, objects for counting (objects that represent ones and objects that represent tens)

TEACHING PAGE 17:
Give the students a set consisting of *6* tens' counters and *3* ones' counters. Ask them what number this represents (63). Tell them to put a set of counters together that represents *35*. Ask them how they would find out how many counters they have altogether. Have them add the ones' counters first (8) and place them to the right. Have them count the tens' counters second (9) and place them to the left. Now ask them to say the new number (98).

Turn to page 17. Compare the illustration at the top of the page to what the students have just done with their counters. Ask the students to look at the first row of problems and say aloud the numbers that are in the ones' place and the numbers that are in the tens' place. Remind them that they should add the ones' place first and the tens' place second. The students may complete this page but their work should be monitored to be sure they are working ones' place first, tens' place second. When the page is completed, have the students read the problems aloud as a group. *Twenty-two plus thirty-one equals fifty-three.*

Add tens to tens.	$+\dfrac{\begin{array}{r}63\\35\end{array}}{98}$	Add ones to ones.
$6 + 3 = 9 \longrightarrow$		$\longleftarrow 3 + 5 = 8$

Write the answer.

22 + 31 53	46 + 41 87	72 + 16 88	34 + 51 85	61 + 18 79	82 + 17 99
72 + 20 92	45 + 30 75	53 + 24 77	32 + 24 56	77 + 11 88	60 + 35 95
80 + 10 90	73 + 25 98	50 + 25 75	75 + 13 88	41 + 42 83	62 + 15 77
17 + 42 59	22 + 51 73	12 + 53 65	83 + 14 97	16 + 22 38	14 + 72 86
43 + 31 74	55 + 33 88	22 + 11 33	17 + 61 78	88 + 11 99	34 + 22 56

page 17 (seventeen)

Page 18: Addition

CONCEPT(S): addition of ones and tens

TEACHER GOAL(S): To teach the children
To review addition facts,
To review columnar addition, and
To add numbers in tens' place.

MATERIALS/MANIPULATIVES:
pencils

TEACHING PAGE 18:

Turn to page 18. Read the directions at the top of the page with the students and review the three types of addition problems on the page. Tell the children they should use the number line only when necessary.

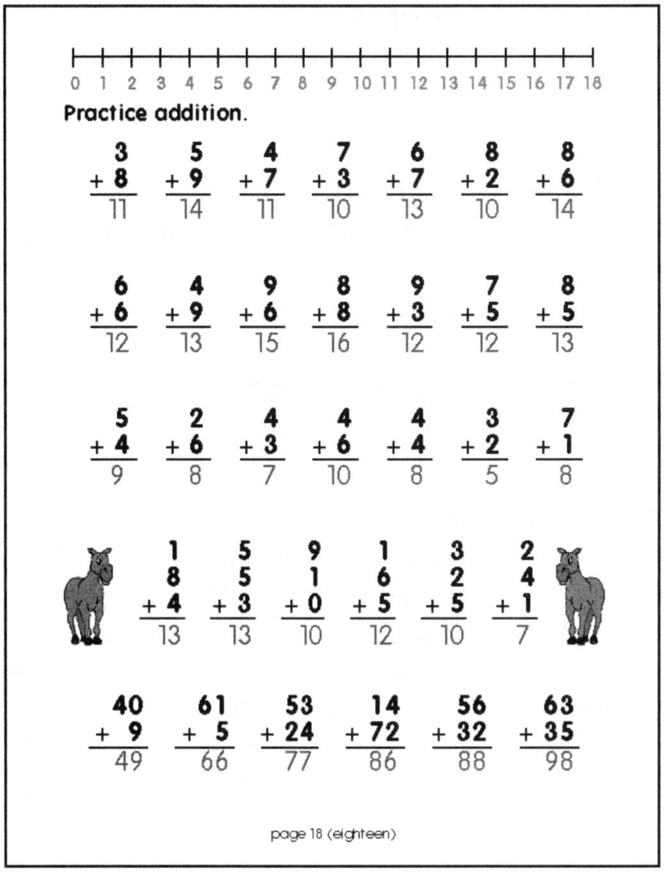

0 1 2 3 4 5 6 7 8 9 10 11 12 13 14 15 16 17 18

Practice addition.

3	5	4	7	6	8	8
+ 8	+ 9	+ 7	+ 3	+ 7	+ 2	+ 6
11	14	11	10	13	10	14

6	4	9	8	9	7	8
+ 6	+ 9	+ 6	+ 8	+ 3	+ 5	+ 5
12	13	15	16	12	12	13

5	2	4	4	4	3	7
+ 4	+ 6	+ 3	+ 6	+ 4	+ 2	+ 1
9	8	7	10	8	5	8

1	5	9	1	3	2
8	5	1	6	2	4
+ 4	+ 3	+ 0	+ 5	+ 5	+ 1
13	13	10	12	10	7

40	61	53	14	56	63
+ 9	+ 5	+ 24	+ 72	+ 32	+ 35
49	66	77	86	88	98

page 18 (eighteen)

Page 19: Number Line

CONCEPT(S): number line to 18

TEACHER GOAL(S): To teach the children
To practice using the number line
to 18, and
To review addition facts to 18.

MATERIALS/MANIPULATIVES:
pencils, objects for counting (ones)

TEACHING PAGE 19:

Turn to page 19, read the title of the page with the students, and review the number line. Read the story with the students and have them identify the things that will be brought to the party. Tell the children to point to 9 on the number line (the number of forks that John is bringing) and count how many more to 18 (9). Have them write a number sentence illustrating the number they started with and how may more they needed to add to reach 18 (*9 + 9 = 18*). Continue in this manner to complete the page. The last question may be answered in several ways. Some students may estimate 2 for the number of children planning the party; others may estimate 18. Discuss with the children why they answered as they did. (For additional practice with number sentences, a similar game may be played with the number 12. Place a set of objects in front of the students and ask how many more to make 12. Let them use the objects or the number line to find the answer. Continue with several combinations to make 12 or any other number up to 18.)

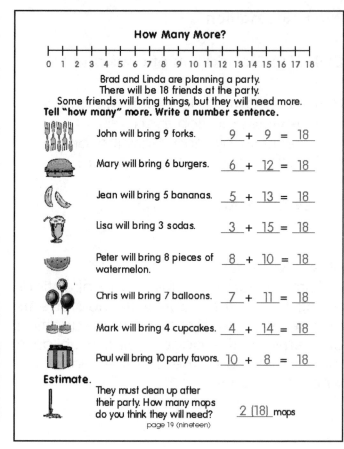

How Many More?

0 1 2 3 4 5 6 7 8 9 10 11 12 13 14 15 16 17 18

Brad and Linda are planning a party.
There will be 18 friends at the party.
Some friends will bring things, but they will need more.
Tell "how many" more. Write a number sentence.

John will bring 9 forks.　　9 + 9 = 18

Mary will bring 6 burgers.　6 + 12 = 18

Jean will bring 5 bananas.　5 + 13 = 18

Lisa will bring 3 sodas.　　3 + 15 = 18

Peter will bring 8 pieces of　8 + 10 = 18
watermelon.

Chris will bring 7 balloons.　7 + 11 = 18

Mark will bring 4 cupcakes.　4 + 14 = 18

Paul will bring 10 party favors.　10 + 8 = 18

Estimate.

They must clean up after
their party. How many mops
do you think they will need?　2 (18) mops
page 19 (nineteen)

Page 20: Place Value

CONCEPT(S): place value for 10's and 1's

TEACHER GOAL(S): To teach the children
To recognize place value for numbers in the tens' place and ones' place.

MATERIALS/MANIPULATIVES:
pencils, crayons, objects for counting (ones and tens)

TEACHING PAGE 20:

Turn to page 20. Point to the first exercise. Ask the students to illustrate the first number (43) using the counters for tens and ones. Ask what the value is of each number symbol. Explain to them that the number 43 can be written showing the value of the number in the tens' place and the number in the ones' place.

Use counters to show how many tens and ones.

$$43 = 4 \text{ tens and } 3 \text{ ones.}$$

Talk about the place value of each number symbol.

$$4 \text{ tens} = 40 \text{ and } 3 \text{ ones} = 3.$$

Write the number showing the place values for tens and ones.

$$43 = 40 + 3$$

Go through the same procedure with the number 74. Then, have the children complete the second exercise.

In the third exercise, reverse the procedure. Have the children start with the place values, convert them to 'how many' tens and 'how many' ones, and then write the number.

Have the children complete the addition problems.

Write how many. Write the value.

43 = _4_ tens + _3_ ones   74 = _7_ tens + _4_ ones
  = _40_ + _3_   = _70_ + _4_

98 = _9_ tens + _8_ ones   21 = _2_ tens + _1_ ones
  = _90_ + _8_   = _20_ + _1_

56 = _5_ tens + _6_ ones   13 = _1_ tens + _3_ ones
  = _50_ + _6_   = _10_ + _3_

Write the number.

70 + 5 = _75_   0 + 3 = _3_

10 + 6 = _16_   40 + 9 = _49_

50 + 0 = _50_   80 + 2 = _82_

Find the sum.

62	37	83	53	71	66
+ 4	+ 2	+ 6	+ 22	+ 13	+ 32
66	39	89	75	84	98

page 20 (twenty)

Page 21: Story Problems

CONCEPT(S): addition and subtraction in story problems

TEACHER GOAL(S): To teach the children To apply the skills they have learned to addition and subtraction story problems.

MATERIALS/MANIPULATIVES:
pencils, paper, crayons, objects for counting (tens and ones)

TEACHING PAGE 21:

Turn to page 21 and read the directions at the top of the page. Ask the students to read the first problem. Students who are able may write the problem in columnar addition, solve the problem, and label the answer. Other students may need to draw pictures of the glasses on paper to form a mental picture of the problem. Students who find it helps to draw the pictures should still complete the steps of columnar addition and labeling the answer. Proceed in this manner to guide the children in solving the remainder of the problems using illustrations or objects for counting wherever helpful.

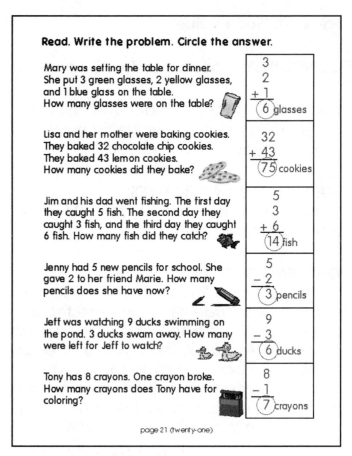

Read. Write the problem. Circle the answer.

Mary was setting the table for dinner. She put 3 green glasses, 2 yellow glasses, and 1 blue glass on the table. How many glasses were on the table?

$$\begin{array}{r} 3 \\ 2 \\ + 1 \\ \hline 6 \end{array}$$ glasses

Lisa and her mother were baking cookies. They baked 32 chocolate chip cookies. They baked 43 lemon cookies. How many cookies did they bake?

$$\begin{array}{r} 32 \\ + 43 \\ \hline 75 \end{array}$$ cookies

Jim and his dad went fishing. The first day they caught 5 fish. The second day they caught 3 fish, and the third day they caught 6 fish. How many fish did they catch?

$$\begin{array}{r} 5 \\ 3 \\ + 6 \\ \hline 14 \end{array}$$ fish

Jenny had 5 new pencils for school. She gave 2 to her friend Marie. How many pencils does she have now?

$$\begin{array}{r} 5 \\ - 2 \\ \hline 3 \end{array}$$ pencils

Jeff was watching 9 ducks swimming on the pond. 3 ducks swam away. How many were left for Jeff to watch?

$$\begin{array}{r} 9 \\ - 3 \\ \hline 6 \end{array}$$ ducks

Tony has 8 crayons. One crayon broke. How many crayons does Tony have for coloring?

$$\begin{array}{r} 8 \\ - 1 \\ \hline 7 \end{array}$$ crayons

page 21 (twenty-one)

Page 22: Oral Directions

CONCEPT(S): responding to oral directions

TEACHER GOAL(S): To teach the children To solve number problems, following oral directions.

MATERIALS/MANIPULATIVES:
pencils

TEACHING PAGE 22:

Turn to page 22. Explain to the students that they must listen carefully and then write the problem. If they do not write a number correctly, the answer to the problem will be wrong. Some leniency may be allowed in spelling for number word problems. When the page is complete, incorrect spelling should be pointed out to the student and the student may be given the opportunity to review.

1) Number facts should be written in number symbols. Students should write the problem and then solve for the answer. Problems must contain signs (+, −) and lines.

Dictate:

$$5 \qquad 12 \qquad 2 \qquad 6$$
$$+7 \qquad -4 \qquad +8 \qquad -3$$

2) The first number sentence should be written in number symbols. The second should be written in number words. Students should write the problem and then solve for the answer.

Dictate: $6 + 7 =$

Seven minus five equals

3) The complete sentence is dictated for equal, not equal, greater than, less than. The student may use number symbols not number words.

Dictate: $7 + 8 = 9 + 6 \qquad 6 - 3 \neq 7 - 5$
$9 < 12 \qquad\qquad 23 > 21$

Listen and Write

Number Facts

5 $+7$ $\overline{12}$	12 -4 $\overline{8}$	2 $+8$ $\overline{10}$	6 -3 $\overline{3}$

Number Sentences

$6 + 7 = 13$

Seven minus five equals two.

Equal (=) / **Not equal (≠)**   **Greater than (>)** / **Less than (<)**

$7 + 8 = 9 + 6$	$9 < 12$
$6 - 3 \neq 7 - 5$	$23 > 21$

Write the number.   **Find the sum.**

twenty-two	24 $+35$ $\overline{59}$	53 $+46$ $\overline{99}$
sixty-four		
thirty-seven		

page 22 (twenty-two)

---

4) The student should write number words.
Dictate: twenty-two sixty-four
thirty-seven

5) Numbers should be written in number symbols to find the sum. An important part of this exercise is to discover whether the students will line the numbers up correctly.

Dictate: $24 \qquad\qquad 53$
$+35 \qquad\qquad +46$

SELF TEST 3:

CONCEPT(S): number words, equal, not equal, greater than, less than

TEACHER GOAL(S): To teach the children To learn to check their progress periodically.

MATERIALS/MANIPULATIVES: pencils

TEACHING PAGE 23:

Turn to page 23. Read the directions with the children. Be sure they understand what they are to do. You may repeat the directions but give no other help. Do not have the children check their own work. Check it as soon as you can and go over it with each child. Show him where he did well and where he needs extra help.
Listen and write dictation:

forty-seven

$\begin{array}{r} 42 \\ + 27 \\ \hline \end{array}$

43 > 39

$7 + 4 \neq 5 + 5$

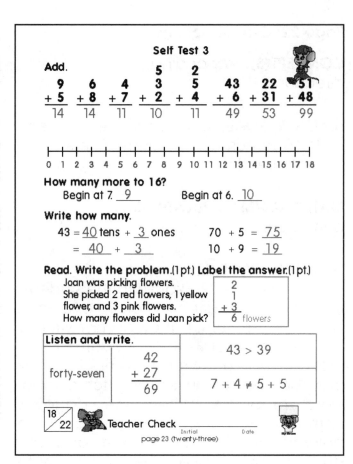

IV. PART FOUR

Pages 24 and 25: Clocks

CONCEPT(S): time to quarter-hour

TEACHER GOAL(S): To teach the children
To tell time to the quarter-hour.

MATERIALS/MANIPULATIVES:
pencils, clock - LIFEPAC 103 page 26 (or similar clock student may use), ruler

TEACHING PAGES 24 and 25:
Review reading the clock to the hour and the half-hour. Give the students several times of the day to the hour and half-hour and have them move the hands on their clocks to show the time given.

Turn to page 24. Point to the first clock. Explain to the students that the *1* on the clock represents five minutes, the *2* represents five more minutes and so on. Have the students count by *5's* starting at the *1* until they reach the *12* or *sixty*. Point to several numbers on the clock and ask the students "how many." (2 represents 10 minutes, 5 represents 25 minutes, 6 represents 30 minutes, and so on). Tell the students to put the hour hand and minute hand on their clocks exactly as they are shown in the first illustration. Talk about how the minute hand on the *3* represents *15* minutes. Point to *5:15* and have the students say it aloud, five fifteen or quarter after 5. Have them practice writing *5:15* on the line indicated. Tell them to use their rulers to divide the clock into four equal parts. Explain to them that when the hand is on the *3*, it is one-fourth or one quarter after *5*. Ask them point to and read *quarter after 5*. Tell them this is another way of saying the time on the clock. Follow the same steps to explain the second illustration (except that the hour hand has now passed the 5). (5:45, five forty-five or quarter till 6).

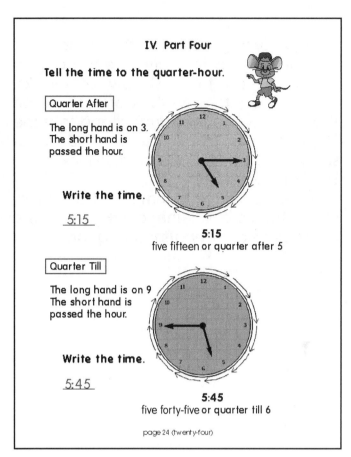

IV. Part Four

Tell the time to the quarter-hour.

Quarter After

The long hand is on 3.
The short hand is passed the hour.

Write the time.

5:15

5:15
five fifteen or quarter after 5

Quarter Till

The long hand is on 9
The short hand is passed the hour.

Write the time.

5:45

5:45
five forty-five or quarter till 6

page 24 (twenty-four)

Turn to page 25. Tell the students to use their clocks to illustrate each one of the times on the first row of clocks. Instruct them to write the correct time on the line. Be sure they understand that the *15* represents *15 minutes* and *45* represents *45 minutes*. Have them count again by *5's* if necessary.

When the page is complete have them read the time on the clocks using the expressions *quarter after* and *quarter till*.

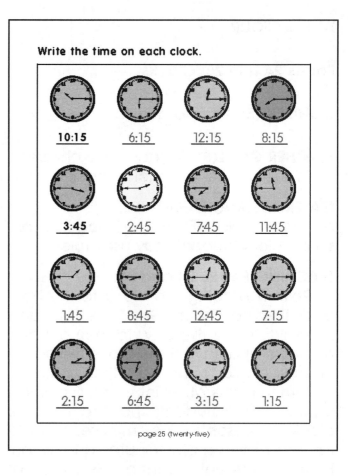

Pages 26 and 27: Money

CONCEPT(S): pennies, nickels, dimes

TEACHER GOAL(S): To teach the children
To count pennies, nickels, and dimes, and
To convert money into coins.

MATERIALS/MANIPULATIVES:
pencils, pennies, nickels, dimes

TEACHING PAGES 26 and 27:
Show the students a dime and ask them to count out the number of pennies that a dime is equal to. Have the students count to 100 by 10's using the dimes. Show the students a nickel and tell them that this coin is equal to five pennies. Have the students count to 100 by 5's using the nickels. Give the students different values and ask them to count out the number of nickels. (25¢, 40¢, 65¢, and so on.)

Turn to page 26. Tell the students to count the pennies for nickels and dimes following the illustration. Have them count by fives for the nickels writing the amount in each box. Then have them fill in the total number of cents. Complete the page using the combinations of coins.

Turn to page 27 and read the instructions at the top of the page with the students. Identify each of the objects and the price tag on the object. Put the pennies, nickels, and dimes in reach of the students. Tell them that they must decide what coins to use to pay for each item. They should then write the number of coins they have used on the line provided. Students may use many different combinations of coins to find their answers. If the total is correct, their answer is correct. It is important that the students experience the use of actual (or play) coins to find the solution to the problem and not simply guess at an answer.

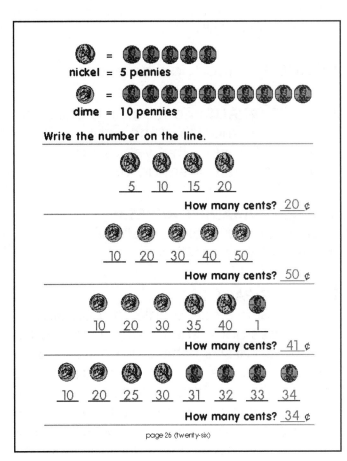

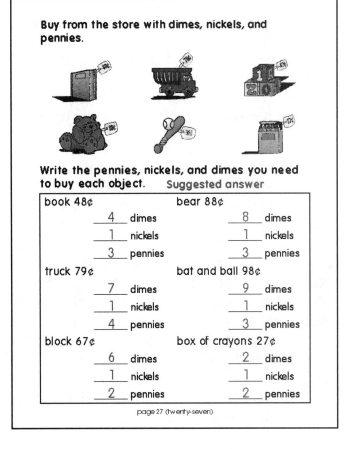

Page 28: Calendar

CONCEPT(S): reading a calendar

TEACHER GOAL(S): To teach the children
To review months of the year,
To read days of the week, and
To write the current date.

MATERIALS/MANIPULATIVES:
pencils, current calendar showing months of year

TEACHING PAGE 28:
There should be a calendar on display in the classroom. The calendar should be reviewed daily so the students become familiar with the months and days of the week. Review this calendar with the students and have them say aloud the months of the year. Ask whether the months all have the same number of days. Review the rhyme:

Thirty days hath September, April, June, and November;
All the rest have thirty-one, excepting February alone:
And that has twenty-eight days clear, and twenty-nine in each leap year.

Turn to page 28. Ask the students the name of the month illustrated on the page. Ask them the number of days in the month and ask if that is what the rhyme tells them. Have them find the same month on their calendars. Discuss whether the first day of the month on their calendars is the same as the illustrated calendar. Students who are able may read and complete the questions. Others may need help in reading. The final question should be completed using the classroom calendar. The answer will be the date on which the student is completing this page.

Calendar
September

Sunday	Monday	Tuesday	Wednesday	Thursday	Friday	Saturday
	1	2	3	4	5	6
7	8	9	10	11	12	13
14	15	16	17	18	19	20
21	22	23	24	25	26	27
28	29	30				

What is the name of the month? _September_
How many days in this month? _30_
How many days in a week? _7_
How many Sundays in this month? _4_
How many Mondays in this month? _5_
Are there more Tuesdays or Wednesdays? _Tuesdays_
Name all the dates for Thursday. _4, 11, 18, 25_
What is the day of the week for September 26? _Friday_

Look at your own calendar. Write today's date.
Teacher check

_____ _____ _____
Day   Month   Date

page 28 (twenty-eight)

SELF TEST 4:

CONCEPT(S): number words, equal, not equal, greater than, less than

TEACHER GOAL(S): To teach the children To learn to check their progress periodically.

MATERIALS/MANIPULATIVES: pencils, clock, pennies, nickels, dimes, current calendar

TEACHING PAGE 29:

Turn to page 29. Read the directions with the children. Be sure they understand what they are to do. You may repeat the directions but give no other help. Do not have the children check their own work. Check it as soon as you can and go over it with each child. Show him where he did well and where he needs extra help. Students may use their clocks and the pennies, nickels, and dimes to complete the test.

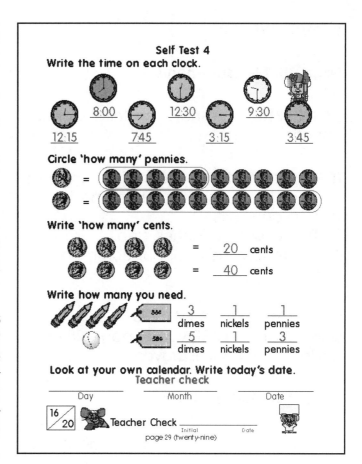

V. PART FIVE

Page 30: Measurements

CONCEPT(S): one-half, one-third, inches, counting

TEACHER GOAL(S): To teach the children
To divide a flat shape into one-third and one-half,
To measure in inches, and
To identify the count of common items.

MATERIALS/MANIPULATIVES:
pencils, ruler - LIFEPAC 101 page 25 or similar six inch ruler, crayons

TEACHING PAGE 30:

Turn to page 30 and review the word *measurements* with the student. Talk about measuring size, amount, length, and weight. Read the directions aloud with the students. They should use their rulers to draw a line through each figure to show $\frac{1}{2}$ or $\frac{1}{4}$. They should then color the part of the figure that represents $\frac{1}{2}$ or $\frac{1}{4}$. Have the children complete the page by measuring the lines and answering the questions.

V. Part Five

Draw a line and color one half ($\frac{1}{2}$) of each flat shape.

Draw a line and color one fourth ($\frac{1}{4}$) of each flat shape.

Measure the line.

_____	_2_ inches
_____	_4_ inches
_____	_1_ inches

How many hours on a clock?	_12_ hours
How many eggs in a dozen?	_12_ eggs
How many nickels in a dime?	_2_ nickels
How many days in a week?	_7_ days

page 30 (thirty)

Page 31: Shapes

CONCEPT(S): flat and solid shapes

TEACHER GOAL(S): To teach the children To use tactile and visual senses to recognize shapes.

MATERIALS/MANIPULATIVES:
pencils, paper bag, spoon, eraser, apple/orange, paper clip, clock/watch, rubber band, block, spool, (or similar items), loose, soft cloth for blindfold, purple and green crayons

TEACHING PAGE 31:

Place the collected items from spoon to spool in the paper bag before starting the day's assignment. Turn to page 31 and read the first set of directions. When the students have identified the flat and solid shapes by circling purple or green, have them say the name of each shape. Talk to the students about the five senses. Tell them that they are going to play a game in which they will need to use their senses of touch, smell, and hearing to identify objects. Go over the list of questions on page 31 and then show the students the bag containing the objects. Tell them that while their eyes are covered, they will pick an item out of the bag. They should feel it, smell it, and listen to it. The item will be put aside (hidden) and they will remove their blindfolds. They must then answer each of the questions on the list about that item. (The teacher may need to assist the students by putting a check mark by the things that apply.) *Name My Shape* requires the student to name the shape (flat or solid) that most closely resembles the selected item. (Some shapes will be difficult to put in a category, but encourage the students to do the best they can.) *What am I* requires the student to find the name of the object from the list of words at the center of the page. *Note: Each item is selected and described before going on to the next item.* When all boxes are filled, give the students the objects in the order in which they took them from the bag. Let them check their own answers to see how well they did. Students may work in teams to complete this page.

page 31 (thirty-one)

Page 32: Addition and Subtraction Facts

CONCEPT(S): addition and subtraction facts

TEACHER GOAL(S): To teach the children To recognize that many facts result in the same answer.

MATERIALS/MANIPULATIVES: pencils, addition and subtraction fact cards

TEACHING PAGE 32:

Review addition and subtraction facts with randomly selected fact cards. Spend some time doing this before beginning the lesson.

Turn to page 32 and read the directions at the top of the page. Assist the students in completing the first group of facts with an answer of 3. Encourage them to select at least one addition and one subtraction fact. For example:

$1 + 2 = 3, \ 4 - 1 = 3, \ 7 - 4 = 3$

The students should be able to complete the page without further assistance. If they are having difficulty, allow them to use the fact cards. Find three fact cards with the answer they are looking for and then copy the fact on page 32.

Write a fact to make the answer true.
Use a fact only once! Add or subtract.

Suggested answer

1	4	7	3	9	2
+ 2	− 1	− 4	+ 4	− 2	+ 5
3	3	3	7	7	7

5	3	10	0	6	9
+ 4	+ 6	− 1	+ 1	− 5	− 8
9	9	9	1	1	1

2	4	9	4	10	9
+ 3	+ 1	− 4	+ 4	− 2	− 1
5	5	5	8	8	8

4	9	8	5	10	6
+ 2	− 3	− 2	+ 5	− 0	+ 4
6	6	6	10	10	10

1	2	7	8	7	1
+ 3	+ 2	− 3	− 6	− 5	+ 1
4	4	4	2	2	2

page 32 (thirty-two)

Page 33: Number Words

CONCEPT(S): number words to ninety-nine

TEACHER GOAL(S): To teach the children To read and write number words to ninety-nine.

MATERIALS/MANIPULATIVES: pencils, paper

TEACHING PAGE 33:

Dictate the words *one* through *ten* to the students and have them write them on paper. Check the spelling immediately and have the students write the words correctly.

Turn to page 33. Have the students read aloud the numbers at the top of the page. Read the directions with the students. Be sure they understand that there are two things for them to do. Tell them to use the numbers at the top of the page for help in reading the words.

Number Words

20-twenty	40-forty	60-sixty	80-eighty
30-thirty	50-fifty	70-seventy	90-ninety

Draw a line to match the number symbols and the number words. Put the number symbols in number order.

53	forty-eight	5	sixty-eight
11	six	68	twenty-one
29	eleven	99	fifty
48	fifty-three	50	five
6	twenty-nine	21	ninety-nine

<u>6   11   29   48   53</u>   <u>5   21   50   68   99</u>

17	eighty-four	72	three
33	thirty-three	87	eighteen
84	two	42	eighty-seven
2	twenty-six	3	forty-two
26	seventeen	18	seventy-two

<u>2   17   26   33   84</u>   <u>3   18   42   72   87</u>

page 33 (thirty-three)

SELF TEST 5:

CONCEPT(S): measurements, shapes, number words

TEACHER GOAL(S): To teach the children To learn to check their progress periodically.

MATERIALS/MANIPULATIVES: pencils, crayons, rulers, addition and subtraction fact cards

TEACHING PAGE 34:

Turn to page 34. Read the directions with the children. Be sure they understand what they are to do. You may repeat the directions but give no other help. Do not have the children check their own work. Check it as soon as you can and go over it with each child. Show him where he did well and where he needs extra help. Students may use fact cards for the last exercise.

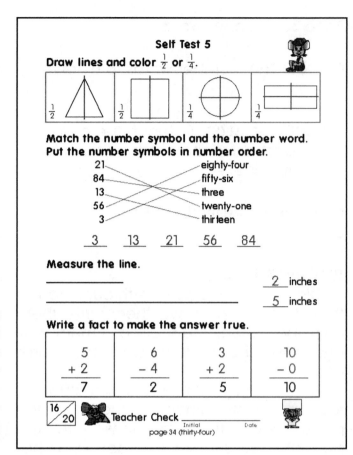

LIFEPAC TEST AND ALTERNATE TEST 101

CONCEPT(S): numbers to 99, addition, subtraction, story problems, time, money, fractions, shapes

TEACHER GOALS: To teach the children
To learn to check their own progress periodically.

MATERIALS/MANIPULATIVES:
pencils, crayons

TEACHING the LIFEPAC TEST:
Administer the test in at least two sessions.
Read all of the directions on each page as the children prepare to do it. Be sure that they understand what they are being asked to do.

LIFEPAC Test page 2
Listen and write.

fifty-six	31	28 > 15
	+ 42	6 + 2 = 4 + 4
	73	

Alternate LIFEPAC Test page 2
Listen and write.

forty-seven	26	49 < 51
	+ 43	3 + 4 = 5 + 2
	69	

Give no help except with directions.
Go over each page with the child as soon as possible after you check it so that he can see where he did well and where he needs more work.
Evaluate the tests and review areas where the children have done poorly. Review the pages and activities that stress the concepts tested.
If necessary, when the children have reviewed sufficiently, administer the Alternate LIFEPAC test. Follow the same procedures as used for the LIFEPAC Test.

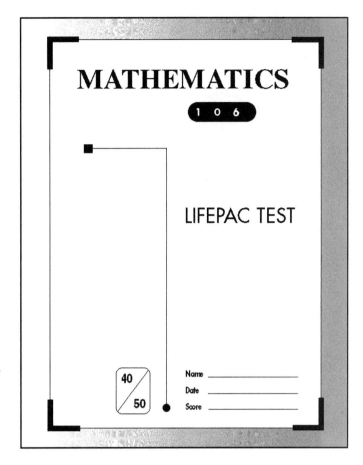

MATHEMATICS 106: LIFEPAC TEST

Write the answer to the facts.

3	7	5	10	5	8
+ 4	+ 8	+ 9	− 6	− 0	− 2
7	15	14	4	5	6

Count by 2's, 5's, 10's. (Each row 1 point)

2, _4_, 6, _8_, _10_, 12, 14,
_5_, 10, _15_, _20_, 25, 30, _35_
10, 20, _30_, _40_, _50_, 60, 70

Circle the odd numbers. (2 points)

(15) 12 (63) (21) 14 18 (29)

Find the sum. Add down. Add up.

11	8	15	12
2	4	7	2
3	1	0	6
+ 6	+ 3	+ 8	+ 4
11	8	15	12

Write the symbol.   Write the words.

twenty-six   _26_   **39** _thirty-nine_
forty-one   _41_   **58** _fifty-eight_

page 1 (one)

Circle (=, ↑).

2 + 6 (= (↑)) 5 + 1

10 − 2 ((=,) ↑) 4 + 4

Circle (>, <).

9 ((>,) <) 5

27 (> (<)) 46

Write a number sentence.

9, 5, 4 $9 - 5 = 4$

ten, two, eight   Ten minus two equals eight.

Write how many.

82 = _8_ tens + _2_ ones   4 tens + 5 ones = _45_

Read. Write the problem. Label the answer. (2 pt.)

Jim read 5 pages in his book on Monday, 2 pages on Tuesday, and 6 pages on Thursday. How many pages did he read altogether?

$$\begin{array}{r} 5 \\ 2 \\ + 6 \\ \hline 13 \text{ pages} \end{array}$$

Listen and write.		
fifty-six	$\begin{array}{r} 31 \\ + 42 \\ \hline 73 \end{array}$	28 > 15
		6 + 2 = 4 + 4

page 2 (two)

Write the time on the clock.

2:45   9:15   10:30

Write how many you need. (1 pt. each amount)

_4_ dimes   _1_ nickels   _1_ pennies

_8_ dimes   _1_ nickels   _2_ pennies

Measure the line.

_____   _3_ inches

Suggested answer

Write a fact to make the answer true.

$\begin{array}{r} 8 \\ -2 \\ \hline 6 \end{array}$	$\begin{array}{r} 7 \\ -4 \\ \hline 3 \end{array}$	$\begin{array}{r} 5 \\ +3 \\ \hline 8 \end{array}$	$\begin{array}{r} 9 \\ +1 \\ \hline 10 \end{array}$

Write the numbers in number order. (2 pts.)

47   23   18   65   91   6

_6_   _18_   _23_   _47_   _65_   _91_

page 3 (three)

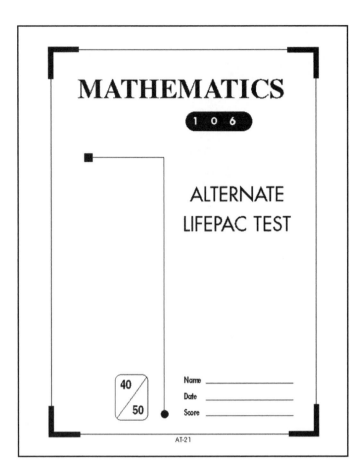

MATHEMATICS

1 0 6

ALTERNATE
LIFEPAC TEST

40 / 50

Name _____
Date _____
Score _____

AT-21

MATHEMATICS 106: Alternate LIFEPAC TEST

Write the answer to the facts.

6	9	4	10	8	7
+ 5	+ 9	+ 8	− 3	− 0	− 5
11	18	12	7	8	2

Count by 2's, 5's, 10's. (Each row 1 point)

2, <u>4</u>, 6, <u>8</u>, <u>10</u>, 12, 14,

<u>5</u>, 10, <u>15</u>, <u>20</u>, 25, 30, <u>35</u>

10, 20, <u>30</u>, <u>40</u>, <u>50</u>, 60, 70

Circle the odd numbers. (2 points)

10 ⟨27⟩ ⟨39⟩ 82 46 ⟨73⟩ ⟨55⟩

Find the sum. Add down. Add up.

8	11	12	11
3	3	5	6
0	6	3	2
+ 5	+ 2	+ 4	+ 3
8	11	12	11

Write the symbol.   **Write the words.**

forty-five   <u>45</u>   62   <u>sixty-two</u>

thirty-three   <u>33</u>   28   <u>twenty-eight</u>

AT-22

Circle (=, ≠).   **Circle (>, <).**

3 + 5 ⟨=⟩ ≠ 4 + 4   12 > ⟨<⟩ 13

8 − 6 ⟨≠⟩ 10 + 8   54 ⟨>⟩ < 43

Write a number sentence.

3, 8, 11   <u>3 + 8 = 11</u>

seven, four three   <u>Seven minus four equals three.</u>

Write how many.

70 = <u>7</u> tens + <u>0</u> ones   3 tens + 6 ones = <u>36</u>

Read. Write the problem. (1 pt.) **Label the answer.** (1 pt.)

Jim read 6 pages in his book on Monday, 3 pages on Tuesday, and 4 pages on Thursday. How many pages did he read altogether?

6
3
+ 4
13 pages

Listen and write.		49 < 51
forty-seven	26 + 43 69	3 + 4 = 5 + 2

AT-23

Write the time on the clock.

2:15   1:30   4:45

Write how many you need.

77¢ → <u>7</u> dimes <u>1</u> nickels <u>2</u> pennies

29¢ → <u>2</u> dimes <u>1</u> nickels <u>4</u> pennies

Measure the line.

————————————————   <u>4</u> inches

Write a fact to make the answer true.

4	7	7	5
+ 4	− 5	− 1	+ 8
8	2	6	13

Write the numbers in number order. (2 pts.)

82   12   65   4   43   29

<u>4</u>   <u>12</u>   <u>29</u>   <u>43</u>   <u>65</u>   <u>82</u>

AT-24

Page 1: Fun With Numbers

CONCEPT(S): purpose of LIFEPAC, objectives

TEACHER GOAL(S): To teach the children
To know what is expected of the student in the LIFEPAC, and
To write first and last names correctly in manuscript.

MATERIALS/MANIPULATIVES:
pencils

TEACHING PAGE 1:

Turn to page 1. Point to the title and the memory verse and read them aloud. Allow time for the children to look through the LIFEPAC. Write the word *OBJECTIVES* on the board and have the children find the word on the page. Explain that the objectives tell the things the students will be expected to do in the LIFEPAC. Read each one and have the children repeat as they run their fingers under the sentence from left to right. Talk about the objectives so that the children will understand what they will be doing. Have each child write his name on the line.

FUN WITH NUMBERS

My name is _____ **Teacher check**

_____

- -

Memory Verse
"If ye abide in me, and my words abide in you, ye shall ask what ye will, and it shall be done unto you."
John 15:7

 **Objectives**

1. I can count and recognize number order to 200.
2. I can learn place value.
3. I can solve story problems.
4. I can subtract number facts to 12.
5. I can learn about A.M., P.M., fractions, shapes and sequence.
6. I can make a chart using data I have gathered.

page 1 (one)

I. PART ONE

Page 2: Numbers to 200

CONCEPT(S): count to 200

TEACHER GOAL(S): To teach the children
To read numbers to 200.

MATERIALS/MANIPULATIVES:
pencils, chart of numbers to 100 -
LIFEPAC 101 page 7 (or any chart that
shows the numbers 1 to 100)

TEACHING PAGE 2:
Review the chart of numbers to *100*
with the students. Have them count aloud
from *80* to *100*.
Turn to page 2. Explain to the children
that when the numbers reach *100*, we
begin counting over again; but, we must
first say *one hundred*. Point to the numbers *90* through *100* at the top of the
page and tell the students to count
aloud. Have the children point to *100* and
say the number that is in the ones' place
(0). Have them say the number that is in
the tens' place (0). Tell them that they
have a new place to learn. It is the
hundreds' place. Have the children point
to the *1* and tell them that this number is
in the *hundreds'* place. Draw the children's attention to the second number on
the second chart (101). Ask them the
same questions. Which number is in the
ones' place? (1) Which number is in the
tens' place? (0) Which number is in the
hundreds' place? (1) Have the children
count aloud from *100* to *200*. Then point to
several numbers at random and have
them tell the number in the ones' place,
the tens' place, and the hundreds' place.

I. Part One

90, 91, 92, 93, 94, 95, 96, 97, 98, 99, 100

Count to 200.

100	101	102	103	104	105	106	107	108	109
110	111	112	113	114	115	116	117	118	119
120	121	122	123	124	125	126	127	128	129
130	131	132	133	134	135	136	137	138	139
140	141	142	143	144	145	146	147	148	149
150	151	152	153	154	155	156	157	158	159
160	161	162	163	164	165	166	167	168	169
170	171	172	173	174	175	176	177	178	179
180	181	182	183	184	185	186	187	188	189
190	191	192	193	194	195	196	197	198	199
200									

page 2 (two)

Page 3: Numbers to 200

CONCEPT(S): write to 200

TEACHER GOAL(S): To teach the children To write numbers to 200.

MATERIALS/MANIPULATIVES:
pencils, cardboard, paste

TEACHING PAGE 3:

It would be helpful to the students to copy or cut out the chart of numbers to *200* on page 2. It should be glued to cardboard to make it more durable.

Turn to page 3 and read the directions with the students. They may use their chart of numbers to complete the page. Help them complete the first line and then monitor them as they complete the page.

Write the number in each box.

100	101	102	103	104	105	106	107	108	109
110	111	112	113	114	115	116	117	118	119
120	121	122	123	124	125	126	127	128	129
130	131	132	133	134	135	136	137	138	139
140	141	142	143	144	145	146	147	148	149
150	151	152	153	154	155	156	157	158	159
160	161	162	163	164	165	166	167	168	169
170	171	172	173	174	175	176	177	178	179
180	181	182	183	184	185	186	187	188	189
190	191	192	193	194	195	196	197	198	199
200									

page 3 (three)

Page 4: Numbers to 200

CONCEPT(S): number order to 200

TEACHER GOAL(S): To teach the children
To recognize number order in
numbers to 200.

MATERIALS/MANIPULATIVES:
pencils, chart of numbers to 200

TEACHING PAGE 4:
Turn to page 4. Read each set of directions aloud with the students. Review *greater than* and *less than* with them. Help them locate the numbers *143* and *162* from the first exercise and *156* and *181* from the second and third exercises on their charts to help them begin. They may use their *chart of numbers to 200* to complete the page.

Write the missing numbers.

143, <u>144</u>, <u>145</u>, 146, <u>147</u>, 148, <u>149</u>
162, <u>163</u>, 164, <u>165</u>, 166, <u>167</u>, 168
101, <u>102</u>, <u>103</u>, <u>104</u>, 105, <u>106</u>, <u>107</u>
192, <u>193</u>, 194, <u>195</u>, <u>196</u>, 197 <u>198</u>
144, <u>145</u>, <u>146</u>, 147 <u>148</u>, <u>149</u>, 150
177 <u>178</u>, 179, <u>180</u>, <u>181</u>, 182, <u>183</u>

Put a circle around each number greater than 156.

(175)	151	111	(199)	(187)	(165)
127	(191)	(157)	129	154	109
130	125	140	(165)	105	123
(182)	(163)	(173)	142	(169)	(180)

Put a circle around each number less than 181.

191	(115)	(159)	(180)	184	193
(122)	(175)	(164)	(153)	192	(124)
187	190	(179)	(110)	(151)	(170)
(116)	196	(145)	200	185	183

page 4 (four)

74

Page 5: Addition Facts to 18

CONCEPT(S): addition facts to 18

TEACHER GOAL(S): To teach the children To estimate how long it will take to complete a page of addition facts.

MATERIALS/MANIPULATIVES:
pencils

TEACHING PAGE 5:

Turn to page 5 and ask the students to tell what kind of problems are on this page (addition facts).

Ask them if they should know most of these facts by now (yes). Draw their attention to the questions at the top and the bottom of the page. Read the two sentences aloud with them. Talk about the word *estimate*. Ask them how long they *estimate* (think) it will take them to complete the page. (Be sure they give a realistic answer.) This should be approached as a challenge without any pressure on the students. Have the students write down an estimated time and then have them tell the time on the clock when they will begin and when they think they will be done. Have the children complete the page and then write how long it took them. Students who become discouraged easily may simply complete the page without estimating the time. *Students who do not do well on this page should spend time on their addition fact cards, reviewing them two to three times a week.*

Addition Facts

Write the answer.   How long do I estimate? ____

Teacher check

6 +4 = 10	4 +9 = 13	8 +3 = 11	7 +8 = 15	7 +6 = 13	8 +8 = 16	9 +4 = 13
6 +8 = 14	9 +2 = 11	8 +6 = 14	6 +6 = 12	9 +8 = 17	3 +8 = 11	5 +5 = 10
9 +3 = 12	8 +4 = 12	8 +9 = 17	5 +6 = 11	4 +7 = 11	3 +7 = 10	9 +9 = 18
9 +5 = 14	7 +7 = 14	6 +9 = 15	6 +7 = 13	7 +5 = 12	9 +6 = 15	3 +9 = 12
4 +8 = 12	9 +7 = 16	8 +5 = 13	7 +9 = 16	9 +1 = 10	8 +2 = 10	4 +6 = 10
5 +7 = 12	5 +8 = 13	7 +3 = 10	8 +7 = 15	7 +4 = 11	6 +5 = 11	5 +9 = 14

Teacher check

How long did I take? ____

page 5 (five)

Page 6: Columnar Addition

CONCEPT(S): addition of three numbers

TEACHER GOAL(S): To teach the children
 To write a number sentence, and
 To add three numbers in a column.

MATERIALS/MANIPULATIVES:
pencils

TEACHING PAGE 6:

Turn to page 6. Read the directions with the students. Ask them what operation is shown in the number sentences (addition) and have them fill in the blanks to complete the sentences. Review the meaning of *sum* (answer to addition problem). Monitor the children's work in the second exercise to be sure they are proceeding correctly. Remind them, in the two digit numbers, to add the ones' place first and the tens' place second.

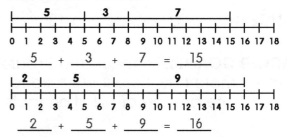

Write a number sentence.

$$\underline{5} + \underline{3} + \underline{7} = \underline{15}$$

$$\underline{2} + \underline{5} + \underline{9} = \underline{16}$$

Write the answer.

2	1	3	2	2	2
1	4	2	1	4	4
+ 2	+ 3	+ 5	+ 5	+ 4	+ 2
5	8	10	8	10	8
6	3	1	2	6	4
4	4	6	8	2	8
+ 2	+ 6	+ 5	+ 3	+ 6	+ 1
12	13	12	13	14	13
60	10	64	32	88	52
+ 4	+ 4	+ 4	+ 3	+ 1	+ 5
64	14	68	35	89	57
34	61	66	50	22	17
+ 51	+ 18	+ 22	+ 25	+ 51	+ 61
85	79	88	75	73	78

page 6 (six)

Page 7: Numbers to 200

CONCEPT(S): count to 200

TEACHER GOAL(S): To teach the children
To draw dot to dot, and
To count to 200.

MATERIALS/MANIPULATIVES:
pencils, chart of numbers to 200, crayons

TEACHING PAGE 7:
Turn to page 7 and read the rhyme with the children. Read the directions and help the students locate numbers *101* and *200* on the picture. Let the children help each other find the numbers and complete the drawing going dot-to-dot. When finished, talk about what part of the cat they drew with the lines. Allow the children to complete the page by coloring the cat.

Number Order to 200

There was a rat,
A small gray rat,
Who had a piece of cheese.
There was a cat,
Named Matt the Cat,
Who often liked to tease.

"I say," the cat said to the rat,
"I'd like some of your cheese."
"You may," the rat said to the cat,
"But first you must say 'please'."

Draw dot-to-dot.
Color.

page 7 (seven)

SELF TEST 1:

CONCEPT(S): count to 200, number facts, addition problems

TEACHER GOAL(S): To teach the children To learn to check their progress periodically.

MATERIALS/MANIPULATIVES: pencils, chart of numbers to 200

TEACHING PAGE 8:

Turn to page 8. Read the directions with the children. Be sure they understand what they are to do. You may repeat the directions but give no other help. Do not have the children check their own work. Check it as soon as you can and go over it with each child. Show him where he did well and where he needs extra help. Students may use the chart of numbers to *200* to complete the test.

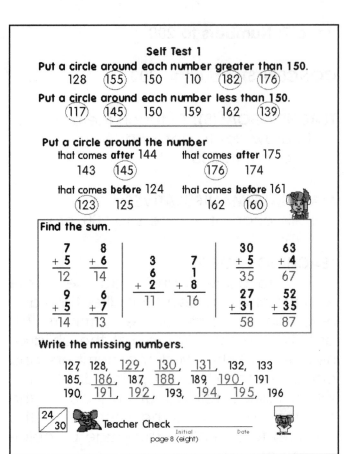

II. PART TWO

Page 9: Ordinal Numbers

CONCEPT(S): ordinal numbers to tenth

TEACHER GOAL(S): To teach the children To tell objects in number order from first to tenth.

MATERIALS/MANIPULATIVES:
pencils, group of ten dissimilar objects

TEACHING PAGE 9:

Place a group of objects in front of the students. Have them describe the objects in order as *first, second, third,* and so on through *tenth.* Point to different objects at random and have the students describe the objects by number order. Have the students rearrange the objects and describe them again in number order.

Turn to page 9. Ask the students to identify each object at the top of the page. Tell them to point to each one in order and say, "first, second, third" continuing to the jack-in-the-box. Read the directions with the students and have them point to the words *first* through *tenth* and say them aloud. Tell the children to read the sentences and write the correct word. Let them write the name of their favorite toy. (It does not have to be from the pictures at the top of the page.)

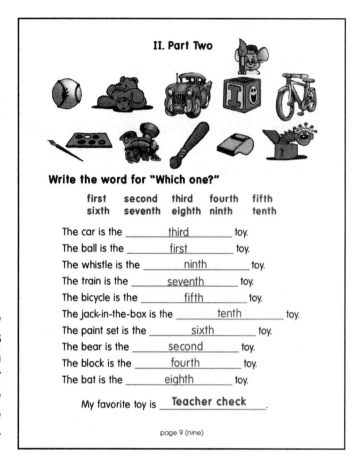

II. Part Two

Write the word for "Which one?"

| first | second | third | fourth | fifth |
| sixth | seventh | eighth | ninth | tenth |

The car is the _____third_____ toy.
The ball is the _____first_____ toy.
The whistle is the _____ninth_____ toy.
The train is the _____seventh_____ toy.
The bicycle is the _____fifth_____ toy.
The jack-in-the-box is the _____tenth_____ toy.
The paint set is the _____sixth_____ toy.
The bear is the _____second_____ toy.
The block is the _____fourth_____ toy.
The bat is the _____eighth_____ toy.

My favorite toy is _____Teacher check_____.

page 9 (nine)

Page 10: Place Value

CONCEPT(S): ones', tens', hundreds' places

TEACHER GOAL(S): To teach the children To identify numbers in ones', tens', and hundreds' places.

MATERIALS/MANIPULATIVES: pencils

TEACHING PAGE 10:

Turn to page 10. Review the meaning of *place value* with the students. Have the students point to the *6* and ask what the place value is of that number (ones). Have them write the *6* on the line in the *1's* column. Ask if there is a number in the tens' place (no), in the hundreds' place (no). Proceed to *78* and ask the same questions and allow the students to fill in the numbers on the lines. Go on to *130*. Review the meaning of *zero* (nothing). Ask the students why, if zero has no value, there is a zero in this number. Write the number *130* on the board and then erase the zero. Ask how that changes the meaning of the number. Tell the children that we call the zero a *place holder*. It is *holding the place* for ones. Use the same procedure for *106*. Ask what position is *being held* in that number. Ask why we do not need a zero in front of *27*. Show that zero in that position does not change the value of the number. Complete the exercise and then read the directions to the second exercise. Ask the children to show two examples in which zero is a *place holder (150, 104)*. Have the students complete the exercise.

Place Value

Write the 100's, 10's, 1's.

		100's		10's		1's
6	=		+		+	6
78	=		+	70	+	8
130	=	100	+	30	+	0
106	=	100	+	0	+	6
27	=		+	20	+	7
145	=	100	+	40	+	5

Write the number.

168	=	100	+	60	+	8
193	=	100	+	90	+	3
150	=	100	+	50	+	0
104	=	100	+	0	+	4
72	=		+	70	+	2
8	=		+		+	8

page 10 (ten)

Page 11: Number Words

CONCEPT(S): number words to ninety-nine

TEACHER GOAL(S): To teach the children To write number words to ninety-nine.

MATERIALS/MANIPULATIVES:
pencils, paper

TEACHING PAGE 11:

Dictate the number words *one* to *ten* to the students and have them write the words on paper. Correct the spelling before going to the next part of the lesson.

Turn to page 11. Read the number words in the boxes aloud with the students. Ask the students if they now have all the number words necessary to write the numbers to ninety-nine. Test the students by having them say a number and then have them point to where the number words are. (*Seventy-nine - seventy* in box on page 11 and *nine* written on students' papers at beginning of lesson.) Continue by letting the students suggest several other numbers and then locating the correct words. Point out to them where hyphens are necessary. Read the directions on the page and let the students complete the page independently.

eleven	fourteen	seventeen
twelve	fifteen	eighteen
thirteen	sixteen	nineteen

Write the number on the line.

eight	8	seventy-eight	78
twelve	12	ninety-nine	99
nineteen	19	fifty	50
five	5	twenty-five	25
fifteen	15	sixty-seven	67
ten	10	eighty-two	82

twenty	fifty	seventy
thirty	sixty	eighty
forty		ninety

Write the word on the line.

16	sixteen	29	twenty-nine
4	four	72	seventy-two
33	thirty-three	96	ninety-six
7	seven	85	eighty-five
11	eleven	67	sixty-seven
58	fifty-eight	44	forty-four

page 11 (eleven)

Page 12: Story Problems with Money

CONCEPT(S): pennies, nickels, dimes

TEACHER GOAL(S): To teach the children
To solve story problems, and
To convert coins to cents using
pennies, nickels, and dimes.

MATERIALS/MANIPULATIVES:
pencils, pennies, nickels, dimes

TEACHING PAGE 12:

Turn to page 12. There are many ways for students to find the correct answers to these problems. Let the children proceed as independently as possible arriving at a solution *using their own methods*. Ask only what they need to know to solve the problem. Encourage counting actual coins to find the answers. For example, in the first problem the students should count out the *27¢* using real (or play) pennies, nickels, and dimes. Some students may do well drawing illustrations (two hamburgers and marking each 45¢). Remind the children that they can count dimes by *10's*, nickels by *5's*, and pennies by *1's* to convert coins to cents.

Dimes, Nickels, and Pennies
Suggested answer

Mary bought 5 suckers and 4 balloons at the store. She gave the clerk 27¢. How many coins did Mary give the clerk?

dimes __2__   nickels __1__   pennies __2__

Joseph and his dad ate lunch at the drive-in. They bought two hamburgers. The hamburgers cost 45¢ each.

How much did Joseph and his dad spend? __90__ cents

Betty had 2 dimes, 3 nickels, and 6 pennies in her pocket. How many cents did Betty have?

dimes __20__ + nickels __15__ + pennies __6__ = __41__ cents
(cents)   _(cents)_   _(cents)_

Betty went to the store and spent 2 nickels and 4 pennies. How many cents does Betty have now?

dimes __20__ + nickels __5__ + pennies __2__ = __27__ cents
(cents)   _(cents)_   _(cents)_

Jack, Paul, and Mark needed a new ball. They each had some nickels. Jack had 6 nickels, Paul had 8 nickels and Mark had 4 nickels.

How many nickels did they have? __18__ nickels
How many cents did they have? __90__ cents
The ball cost 87 cents at the store. Did they have enough money? __yes__

page 12 (twelve)

Page 13: Subtraction Facts

CONCEPT(S): subtraction facts to 9

TEACHER GOAL(S): To teach the children To review subtraction facts to 9.

MATERIALS/MANIPULATIVES: pencils

TEACHING PAGE 13:

Turn to page 13. Review subtraction number facts on the number line. Have the students fill in the blanks to complete the number sentences. Read the next two sets of instructions with the students and have them complete the page independently. *Students having difficulty on this page should review their subtraction facts through 9's two to three times a week.*

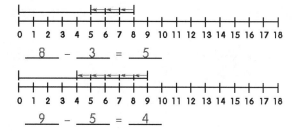

Write a number sentence.

8 – 3 = 5

9 – 5 = 4

Write the answer.

7	8	4	6
−3	−4	−1	−3
4	4	3	3

6 – 4 = 2

9	7	2	5
−6	−5	−1	−0
3	2	1	5

9 – 1 = 8

3 – 3 = 0

8 – 5 = 3

Write the word on the line.

Four – two = two . | Two minus two equals zero .
Eight – six = two . | Nine minus six equals three .
Five – zero = five . | Three minus one equals two .
Seven – three = four . | Eight minus four equals four .

page 13 (thirteen)

Page 14: Count to 100

CONCEPT(S): count to 100, addition facts

TEACHER GOAL(S): To teach the children
To review addition facts, and
To review counting to 100.

MATERIALS/MANIPULATIVES:
pencils, crayons

TEACHING PAGE 14:
Turn to page 14. This may be played as a race between students to see who finishes first or simply used as an exercise in addition. The students begin at the word *Go*. They complete the first fact (5), count along the number line to *5* and color the *5* the same color as the answer box. They complete the second fact (9) and continue from *5* along the number line to *14*. They color the *14* the same color as the answer box for *9*. They proceed in this manner until they have completed all fourteen facts. (They should be at *92*.) They must write a fact (any fact equal to 8) to get them to *100*. They may then circle the answer to the question, "Are you a turtle (slow) or a hare (fast)?" Allow the students to complete this page with as little help as possible. Do not give them the number *92* in advance. If the missing fact has an incorrect sum, review it *after the page is completed.* Point out how important it is to be careful as well as fast. One incorrect answer can make the final answer wrong.

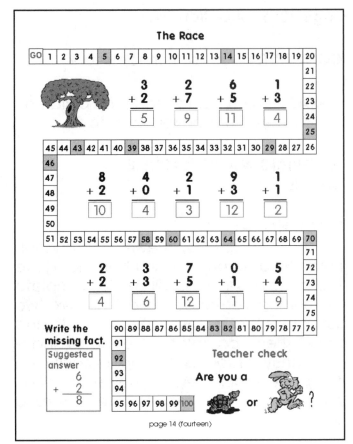

SELF TEST 2:

CONCEPT(S): ordinal numbers, place value, subtraction facts, story problems

TEACHER GOAL(S): To teach the children To learn to check their progress periodically.

MATERIALS/MANIPULATIVES: pencils, pennies, nickels, dimes

TEACHING PAGE 15:

Turn to page 15. Read the directions with the children. Be sure they understand what they are to do. You may repeat the directions but give no other help. Do not have the children check their own work. Check it as soon as you can and go over it with each child. Show him where he did well and where he needs extra help. Students may use pennies, nickels, and dimes to solve the story problem.

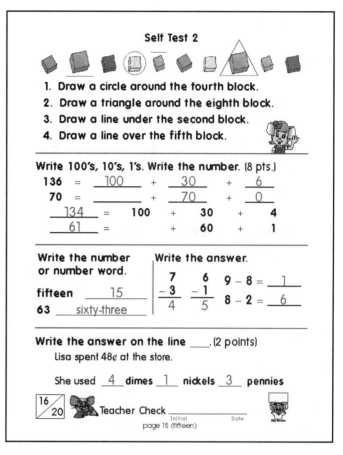

Self Test 2

1. Draw a circle around the fourth block.
2. Draw a triangle around the eighth block.
3. Draw a line under the second block.
4. Draw a line over the fifth block.

Write 100's, 10's, 1's. Write the number. (8 pts.)

136	=	100	+	30	+	6
70	=		+	70	+	0
134	=	100	+	30	+	4
61	=		+	60	+	1

Write the number or number word.

fifteen ___15___

63 ___sixty-three___

Write the answer.

$\begin{array}{r} 7 \\ -3 \\ \hline 4 \end{array}$ $\begin{array}{r} 6 \\ -1 \\ \hline 5 \end{array}$

$9 - 8 = 1$

$8 - 2 = 6$

Write the answer on the line ___. (2 points)

Lisa spent 48¢ at the store.

She used __4__ dimes __1__ nickels __3__ pennies

16/20 Teacher Check _____

Initial   Date

page 15 (fifteen)

III. PART THREE

Page 16: Subtraction Facts

CONCEPT(S): subtraction facts to 12's

TEACHER GOAL(S): To teach the children
To learn subtraction facts for 10's, 11's, 12's.

MATERIALS/MANIPULATIVES:
pencils, new subtraction fact cards on 2 inch by 3 inch cardboard using facts as shown on page 16 - two sets may be made with and without the answer on the back of the card

TEACHING PAGE 16:
Turn to page 16. Review subtraction on the number line with the students. Have the children locate *10* on the number line. Then tell them to count back *zero* places. Tell them to write the answer to *10* minus *0* in the first box. Continue in this manner. Be sure the students are writing the correct answer for each problem. *Writing an incorrect answer will reinforce it in their minds.* Review the subtraction facts (10-12) with the students using the new fact cards.

III. Part Three

0 1 2 3 4 5 6 7 8 9 10 11 12 13 14 15 16 17 18

Subtract 10's, 11's, 12's.

10 −0 = 10	10 −1 = 9	10 −2 = 8	10 −3 = 7	10 −4 = 6	10 −5 = 5
10 −6 = 4	10 −7 = 3	10 −8 = 2	10 −9 = 1	10 −10 = 0	11 −0 = 11
11 −1 = 10	11 −2 = 9	11 −3 = 8	11 −4 = 7	11 −5 = 6	11 −6 = 5
11 −7 = 4	11 −8 = 3	11 −9 = 2	11 −10 = 1	11 −11 = 0	12 −0 = 12
12 −1 = 11	12 −2 = 10	12 −3 = 9	12 −4 = 8	12 −5 = 7	12 −6 = 6
12 −7 = 5	12 −8 = 4	12 −9 = 3	12 −10 = 2	12 −11 = 1	12 −12 = 0

page 16 (sixteen)

Page 17: Subtraction Facts

CONCEPT(S): subtraction facts to 12's

TEACHER GOAL(S): To teach the children
To learn subtraction facts for 10's,
11's, 12's.

MATERIALS/MANIPULATIVES:
pencils, subtraction fact cards for 10's,
11's, 12's, objects for counting

TEACHING PAGE 17:
Review the subtraction fact cards in
order for *10's, 11's,* and *12's.*

Turn to page 17 and review the use of
the number line for subtraction problems.
Read the directions with the students.
Encourage them to use the number line
to complete the page. Some students
may respond better using the objects for
counting. Have them make sets of *10*
objects, *11* objects, and *12* objects. They
will then have the sets ready to use as
they complete each problem.

Write the answer.

10 − 9 1	12 − 6 6	11 − 7 4	10 − 2 8	11 − 9 2	10 − 1 9	12 − 8 4
11 − 8 3	10 − 4 6	12 − 8 4	10 − 7 3	12 − 3 9	12 − 6 6	10 − 3 7
12 − 5 7	11 − 3 8	10 − 5 5	12 − 9 3	10 − 3 7	11 − 6 5	12 − 9 3
12 − 4 8	11 − 2 9	10 − 6 4	11 − 5 6	10 − 9 1	11 − 4 7	10 − 8 2
10 − 7 3	12 − 7 5	11 − 9 2	10 − 5 5	11 − 8 3	12 − 3 9	11 − 3 8
12 − 7 5	10 − 2 8	10 − 1 9	11 − 7 4	12 − 6 6	11 − 9 2	11 − 6 5

page 17 (seventeen)

Page 18: Operation Symbols

CONCEPT(S): operation symbols for =, ≠

TEACHER GOAL(S): To teach the children
To use operation symbols for equal (=)
and not equal (≠).

MATERIALS/MANIPULATIVES:
pencils, subtraction fact cards for 10's,
11's, 12's, objects for counting

TEACHING PAGE 18:
Turn to page 18 and review the meaning and symbols for equal and not equal. Read the directions with the students and point out that some problems are in number symbols and some problems are in number words. Tell them they may use the number line or objects for counting to help them answer the questions.

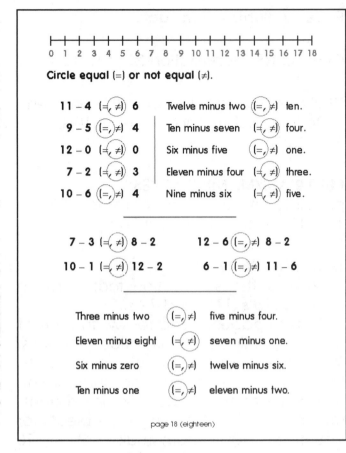

page 18 (eighteen)

Page 19: Addition and Subtraction Facts

CONCEPT(S): addition facts to 18, subtraction facts to 12

TEACHER GOAL(S): To teach the children To review addition facts to 18 and subtraction facts to 12.

MATERIALS/MANIPULATIVES: pencils, subtraction fact cards for 10's, 11's, 12's, objects for counting

TEACHING PAGE 19:

Review subtraction fact cards in order for *10's, 11's, and 12's.*

Turn to page 19. Ask the children what is on this page (addition and subtraction facts). Read the directions with them and have them complete the page independently. They may use the number line or objects for counting.

Add or subtract.

3 $+5$	5 $+5$	9 $+2$	2 $+4$	7 $+3$	6 $+6$	3 $+2$
8	10	11	6	10	12	5

5 $+9$	8 $+8$	6 $+1$	3 $+8$	8 $+7$	7 $+4$	6 $+2$
14	16	7	11	15	11	8

9 $+3$	5 $+8$	7 $+7$	9 $+1$	7 $+5$	4 $+4$	8 $+4$
12	13	14	10	12	8	12

4 -1	10 -3	11 -2	12 -8	9 -3	12 -7	10 -6
3	7	9	4	6	5	4

7 -7	11 -7	10 -5	9 -6	8 -7	11 -9	11 -4
0	4	5	3	1	2	7

10 -7	10 -8	12 -4	11 -8	8 -5	9 -8	12 -9
3	2	8	3	3	1	3

page 19 (nineteen)

Page 20: What Makes Twelve?

CONCEPT(S): twelve in a dozen, twelve hours on the clock

TEACHER GOAL(S): To teach the children To count 12 in a dozen, and 12 hours on the clock.

MATERIALS/MANIPULATIVES: pencils, twelve objects for counting, clock - LIFEPAC 103 - page 26 (or similar clock for student use)

TEACHING PAGE 20:

Turn to page 20. Talk to the children about "how many" in a dozen. Let them use objects for counting as they read the story problems and write the answers. Talk to the children about the number of hours on a clock. Let them use their own clocks to illustrate the story problems. Let them count the hours on the clock and then write the answers.

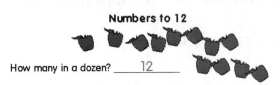

Numbers to 12

How many in a dozen? ___12___

Jane bought a dozen pieces of candy at the store. She ate 4 before she returned home. How many pieces of candy did Jane have left? _8 pieces of candy_

Matthew's mother bought a dozen eggs at the store. When she arrived home 2 were broken. How many eggs did she have left? _10 eggs_

Sally was asked to bake a dozen cookies for her friend's party. There were 8 on the first cookie tray. How many more did she need to bake? _4 cookies_

How many hours on a clock? ___12___

Jessie left the house at 9:00 o'clock. Her mother told her to be home by 12:00 o'clock. How many hours could Jessie be gone? _3 hours_

Jack's dad goes to work at 6:00 o'clock and comes home for lunch at 12:00 o'clock. How many hours does Jack's dad spend at work before lunch? _6 hours_

Roger needed to be home by 12:00 o'clock. He wanted to be able to play with his friend for 2 hours. What time could Roger start playing? _10:00 o'clock_

page 20 (twenty)

Page 21: Number Order

CONCEPT(S): number order to 200

TEACHER GOAL(S): To teach the children
To write numbers before, after, and between,
To find numbers greater than and less than, and
To write numbers in number order.

MATERIALS/MANIPULATIVES:
pencils, chart of numbers to 200

TEACHING PAGE 21:

Turn to page 21 and read the directions with the children. Be sure they understand the directions to each exercise. Explain that there are two steps to the last exercise. The students may use the chart of numbers to *200* to help them complete the page.

Write the number

before and after.

130	131	132		48	49	50
51	52	53		182	183	184
6	7	8		16	17	18

between.

59	60	61		138	139	140
172	173	174		99	100	101
27	28	29		12	13	14

Circle the numbers

greater than 143.

129   (146)   (182)   (176)   101

less than 159.

(129)   (146)   182   176   (101)

Write the answer. Put answers in number order.

$$\begin{array}{ccccc} 6 & 12 & 8 & 11 & 2 \\ +3 & -7 & +5 & -3 & +9 \\ \hline 9 & 5 & 13 & 8 & 11 \end{array}$$

   5   8   9   11   13

page 21 (twenty-one)

SELF TEST 3:

CONCEPT(S): subtraction facts to 12, equal, not equal, numbers to 200, the number 12

TEACHER GOAL(S): To teach the children To learn to check their progress periodically.

MATERIALS/MANIPULATIVES: pencils, chart of numbers to 200

TEACHING PAGE 22:

Turn to page 22. Read the directions with the children. Be sure they understand what they are to do. You may repeat the directions but give no other help. Do not have the children check their own work. Check it as soon as you can and go over it with each child. Show him where he did well and where he needs extra help. Students may use the chart of numbers to *200* to complete the test.

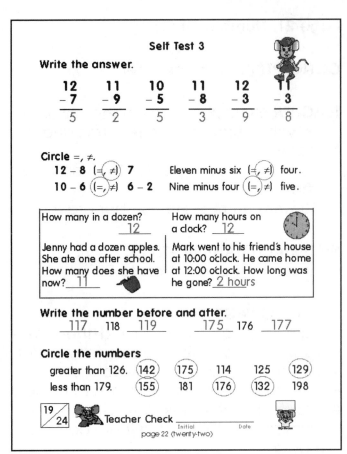

IV. PART FOUR

Page 23: Time to the Quarter-hour

CONCEPT(S): time to the quarter-hour

TEACHER GOAL(S): To teach the children To tell time to the quarter-hour.

MATERIALS/MANIPULATIVES:
pencils, clock - LIFEPAC 103 - page 26 (or similar clock for student use)

TEACHING PAGE 23:
Review telling time using the students' clocks. Have them practice moving the hands on the clocks to tell time to different hours and then have them say the time aloud (six o'clock, nine o'clock, and so on). Ask the children how many minutes in an hour. Discuss that each number on the clock stands for *five* minutes and have the children count by *5's* around the clock. Practice the half-hour (five thirty, ten thirty and so on). Have the children move the hands of the clock to show *fifteen* minutes after the hour and *forty-five* minutes after the hour. Ask them to say the time aloud. Tell them that the clock may be divided into four parts - at the *3, 6, 9,* and *12.* Talk about *5:15* meaning the same as quarter after and *5:45* meaning the same as quarter till.

Turn to page 23 and read the directions with the students. Tell the children that they should not use the expressions *quarter after* and *quarter till* when writing the time. They should use *5:15* and *5:45.* Have the children complete the page but monitor their work. If they have difficulty, remind them to count by *5's.*

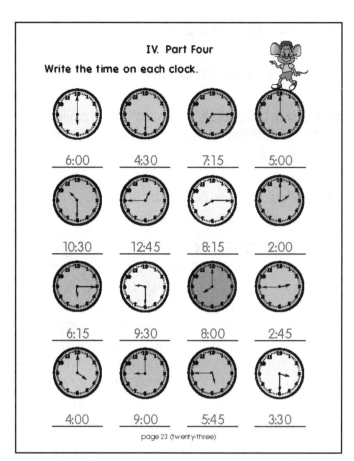

IV. Part Four

Write the time on each clock.

6:00	4:30	7:15	5:00
10:30	12:45	8:15	2:00
6:15	9:30	8:00	2:45
4:00	9:00	5:45	3:30

page 23 (twenty-three)

Page 24: a.m. and p.m.

CONCEPT(S): a.m. and p.m.

TEACHER GOAL(S): To teach the children To understand the meaning of a.m. and p.m.

MATERIALS/MANIPULATIVES:
pencils, clocks for student use

TEACHING PAGE 24:

Talk to the children about morning, afternoon, evening, and night. Ask the children the number of hours in a day (24). Ask how many hours on a clock (12). Ask how many times the hour hand on a clock must go around for a full day (twice). Have the children move the hands on the clock to 12:00 o'clock. Tell them that this is the beginning of the day. Talk to them about what they are doing at midnight. Have them move the hands forward on the clock and discuss with them what they might be doing at various times (waking up, eating breakfast, going to school). Have the children stop the hands when they reach 12:00 o'clock, and ask them what we call this time (noon). Tell them that we describe the first time the hour hand goes around as a.m. Repeat the same steps as the children move the hands around the clock until they reach 12:00 o'clock again. Tell them that we describe the second time the hour hand goes around as p.m.

Turn to page 24. Point to the clocks, and ask the children to write the times shown on the clocks. Ask them to say the times aloud. Read the directions to the children. Explain to them that they may write just a.m. or just p.m., but sometimes they may want to write both. When the children have completed the page, discuss their answers with them.

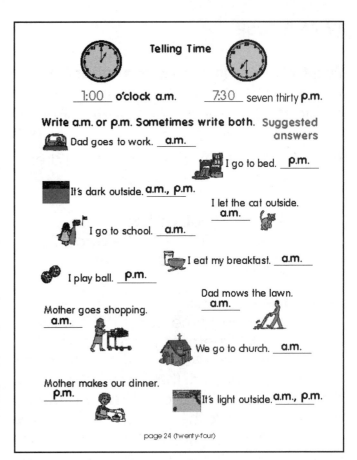

Page 25: Flat Shapes and Fractions

CONCEPT(S): flat shapes, fractions to one-sixth

TEACHER GOAL(S): To teach the children
To write a fraction, and
To recognize the names of flat shapes.

MATERIALS/MANIPULATIVES:
pencils

TEACHING PAGE 25:
Turn to page 25 and read the title with the children. Ask them to draw lines matching the shapes with the names of the shapes. Read the next direction. Tell the children that they will write a fraction to describe the shaded part of each shape. Begin with the triangle. Ask the students how many numbers in a fraction (2) and what separates the numbers (a line). Have the children draw a line. Have them count the parts the triangle is divided into (3) and write that number below the line. Ask how many parts are shaded (1). Tell the children to write that number above the line. Ask them to say the fraction aloud (one-third). Continue in this manner for the circle, the rectangle, and the square. Read the next direction with the students. Ask them how many sets the twelve balls represent (1). Have them begin by drawing a line for the fraction. Ask how many parts the set is divided into (12). Ask where they will write this number (below the line). Ask how many parts are shown outside the set (4). Ask where they will write this number (above the line). Have them say the fraction aloud (four-twelfths). Continue following these steps to complete the exercise.

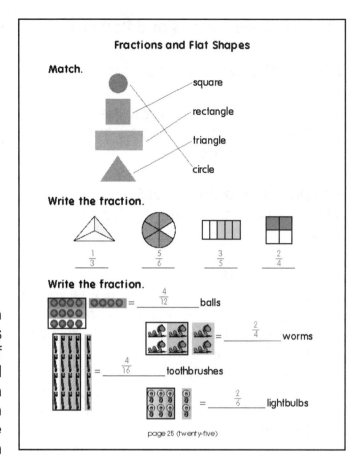

Note: Fractions represent a ratio between two numbers. For example: In the first set on page 25, $\frac{4}{12}$ of the set of balls could also be written as $\frac{1}{3}$ of the set of balls. Stress to the students that in these problems, the sets are divided into parts of 12 balls, 16 brushes, 4 worms, and 6 light bulbs. This means that the objects outside the set represent $\frac{4}{12}$ of the set of balls, $\frac{4}{16}$ of the set of brushes, $\frac{2}{4}$ of the set of worms, and $\frac{2}{6}$ of the set of light bulbs. Have the students circle each object in the set to reinforce the concept of the number of parts, if necessary.

Page 26: Solid Shapes and Patterns

CONCEPT(S): solid shapes and patterns

TEACHER GOAL(S): To teach the children
To recognize solid shapes, and
To recognize patterns.

MATERIALS/MANIPULATIVES:
pencils, objects of different size and shape

TEACHING PAGE 26:
Spend some time with the students creating patterns using objects of different sizes and shapes. Use patterns such as: big - little, up - down, 1-2-1-2, and so on. Have the students tell what comes next. Let the students create patterns and the teacher or other students tell what comes next.

Turn to page 26. Read the directions with the students. Have them say the names of the solid shapes aloud. Students should be able to complete this page independently.

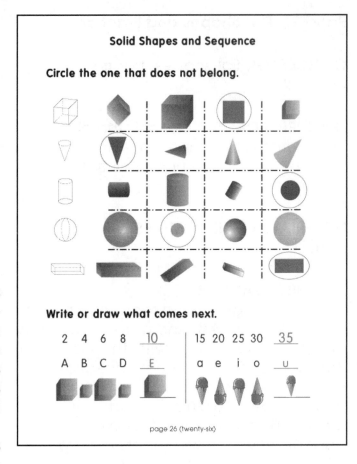

Solid Shapes and Sequence

Circle the one that does not belong.

Write or draw what comes next.

2 4 6 8 <u>10</u> | 15 20 25 30 <u>35</u>

A B C D <u>E</u> | a e i o <u>u</u>

page 26 (twenty-six)

SELF TEST 4:

CONCEPT(S): time, shapes, fractions, and patterns

TEACHER GOAL(S): To teach the children To learn to check their progress periodically.

MATERIALS/MANIPULATIVES: pencils

TEACHING PAGE 27:

Turn to page 27. Read the directions with the children. Be sure they understand what they are to do. You may repeat the directions but give no other help. Do not have the children check their own work. Check it as soon as you can and go over it with each child. Show him where he did well and where he needs extra help.

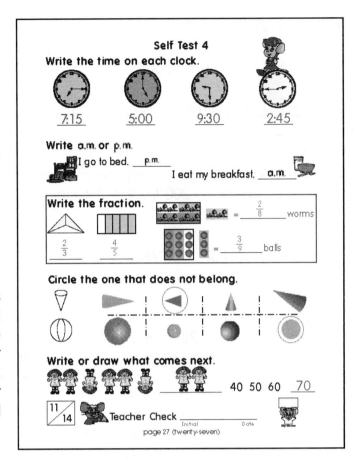

V. PART FIVE

Page 28: Skip Count to 200

CONCEPT(S): count to 200 by 2's, 5's, and 10's

TEACHER GOAL(S): To teach the children To skip count to 200 by 2's, 5's, and 10's.

MATERIALS/MANIPULATIVES: pencils, purple crayon

TEACHING PAGE 28:

Turn to page 28. Point to the two charts and ask the children what is alike and what is different about them. Emphasize the similarities. Have the children count aloud by *2's*, *5's*, and *10's* on the chart to *100*. Have them point to the second chart and count by *1's*, *2's*, *5's*, and *10's*. Read the directions with the children and allow them to complete the page.

page 28 (twenty-eight)

Page 29: Operation Symbols

CONCEPT(S): operation symbols

TEACHER GOAL(S): To teach the children
To use operation symbols (+, −, =, ≠, >, <).

MATERIALS/MANIPULATIVES:
pencils, subtraction fact cards for 10's, 11's, 12's, number symbol cards for 10, 11, 12

TEACHING PAGE 29:
Turn to page 29 and read the title with the students. Explain to them that the symbols at the top of the page are called *operation symbols*. Read the directions for each exercise and be sure the students understand what they are to do. Allow the children time to complete the page. When the page is finished, select a group of subtraction fact cards and number symbol cards. Play a game of concentration with the students.

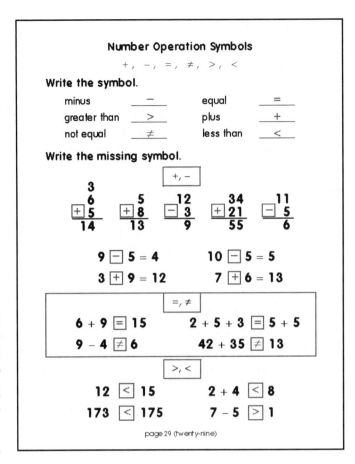

Page 30: Estimation

CONCEPT(S): estimation

TEACHER GOAL(S): To teach the children To estimate amounts and then verify the estimation using tactile methods.

MATERIALS/MANIPULATIVES:
pencils, 25 to 30 objects - dried beans are ideal, container to hold beans - large enough and safe enough for students to place hand into

TEACHING PAGE 30:
Discuss the meaning of *estimation* with the students. Have them estimate the time of day (without looking at the clock), how long it takes to walk from home to a friend's house, and similar things. Explain that estimation is more than guessing. We use facts to find an estimation that is reasonable. Place the container with the beans in front of the students and explain that they will do a lesson in estimation today.

Turn to page 30. Point to the jar and explain that the jar is like their container. Read the instructions with the students and have them look at the first exercise. Tell them, with their eyes closed, they should draw *3* beans from the container. Tell them they cannot count the beans (1-2-3); they must pick as close to *3* as they can. Next, they should actually count the number of beans they picked and write the number down on the line. Finally, they should write an operation symbol in the box to compare the number they were suppose to draw to the number they did draw (=, >, <). Have the students proceed in this manner through exercise 6. Ask the students to tell how many equal signs they had. (How many times they picked the number they were suppose to pick.) Go to the next set of six exercises. Tell the

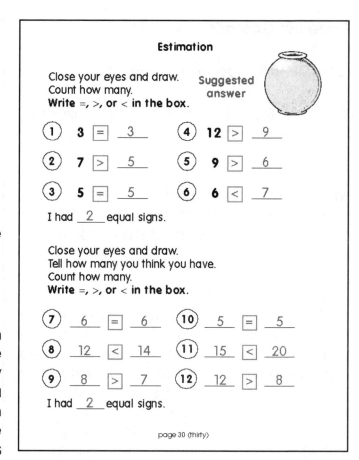

students to close their eyes. This time have them draw and then estimate the number they have drawn. (Allow the children time to feel *but not* count the number of beans.) Have them write this number on the first line and then count the number they have actually drawn. Complete the exercise by entering the correct number symbol in the box. Continue in this manner to complete the page.

Page 31: Charts

CONCEPT(S): charting data

TEACHER GOAL(S): To teach the children
To put numbers on a chart.

MATERIALS/MANIPULATIVES:
pencils, crayons, ruler

TEACHING PAGE 31:

Turn to page 31 and read the title at the top of the page. Help the children identify the two charts on the page. Show them the numbers along the side. Tell them that these are the same as the numbers at the bottom of page 30. Let them count by *5's* along the bottom of each chart. On the first chart, have the children color in the number of beans that they estimated they had drawn in exercises 7 through 12. Tell them to use a different color crayon for each line. On the second chart, have the children color in the number of beans that they actually drew in exercises 7 through 12. Tell them to use the corresponding (to the first chart) color crayon for each line. Where students drew a number such as *7*, have them identify that *7* as between *5* and *10* and color to that point. Tell them to compare the two charts to see how closely the estimation came to the actual count.

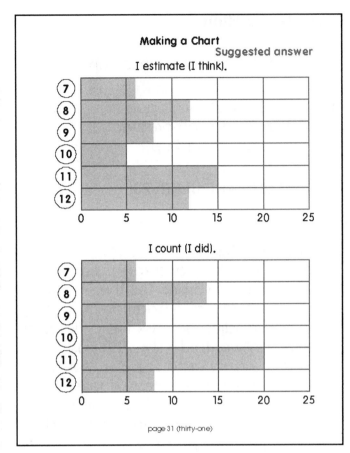

101

Page 32: Addition

CONCEPT(S): checking addition

TEACHER GOAL(S): To teach the children
To check addition problems by
adding down and up.

MATERIALS/MANIPULATIVES:
pencils

TEACHING PAGE 32:
Turn to page 32 and read the instructions with the children. Tell them that we can check answers in addition by adding down and adding up. Ask the children to write the answer to *3 + 2 =*. Then ask them to write the answer to *2 + 3 =*. Ask if the answer is the same. Have the children complete the page in this manner. Be sure they are adding both down and up and not just copying from one box to another. Have the students do the work orally, if necessary.

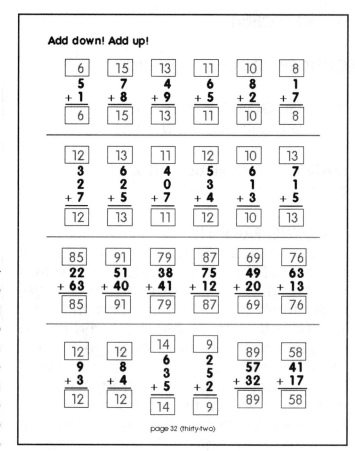

Page 33: Big and Little Numbers

CONCEPT(S): place value and number size

TEACHER GOAL(S): To teach the children To understand place value and number size.

MATERIALS/MANIPULATIVES: pencils, number symbol cards 0 through 9 (2 inch by 3 inch cards with numbers written on cards)

TEACHING PAGE 33:

Play a game with the students with the number symbol cards. For example: Select the *1* card and the *5* card. Put them together to form the number *15* and then the number *51*. Ask the students which is the *big* number and which is the *little* number. Select different symbol cards at random and have the students place them together to make a *big* number and a *little* number. Talk to the children about the fact that the *larger* number is always in the tens' place to form the *big* number.

Turn to page 33. Read the directions on the page. The students should be able to complete the page independently.

Big and Little

Use two number symbols.
Write a Big **number. Write a** Little **number.**

	Big	Little
3, 5	⑤3	③5
7, 4	⑦4	④7
0, 6	⑥0	6
2, 9	⑨2	②9
8, 3	⑧3	③8

Circle the numbers in the 10's place for the Big and Little numbers.

The Big number always has the largest number symbol in the (⑩'s , 1's) **place.**

Use number symbols or number words.
Write number facts.

6, 5, 11 __6 + 5 = 11__   4, 3, 7 __4 + 3 = 7__
12, 3, 9 __12 − 3 = 9__   8, 2, 6 __8 − 2 = 6__
four, two, six ____Four plus two equals six.____
eight, five, three ____Eight minus five equals three.____

page 33 (thirty-three)

SELF TEST 5:

CONCEPT(S): skip counting to 200, operation symbols, estimation, big and little numbers.

TEACHER GOAL(S): To teach the children
To learn to check their progress periodically.

MATERIALS/MANIPULATIVES:
pencils, crayons

TEACHING PAGE 34:

Turn to page 34. Read the directions with the children. Be sure they understand what they are to do. You may repeat the directions but give no other help. Do not have the children check their own work. Check it as soon as you can and go over it with each child. Show him where he did well and where he needs extra help. In the first exercise, count two points for completing the 2's, two points for completing the 5's, and two points for completing the 10's.

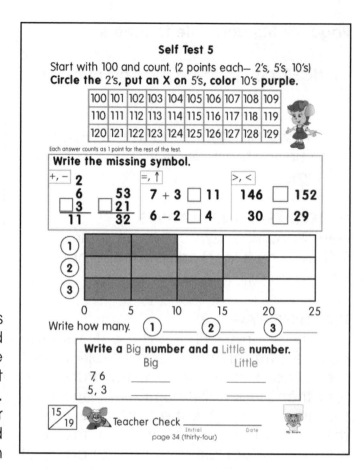

LIFEPAC TEST AND ALTERNATE TEST 107

CONCEPT(S): numbers to 200, add to 18, subtract to 12, fractions, shapes, operation signs

TEACHER GOALS: To teach the children
   To learn to check their own progress periodically.

MATERIALS/MANIPULATIVES:
pencils

TEACHING the LIFEPAC TEST:
   Administer the test in at least two sessions.
   Read all of the directions on each page as the children prepare to do it. Be sure that they understand what they are being asked to do.
   Give no help except with directions.
   Go over each page with the child as soon as possible after you check it so that he can see where he did well and where he needs more work.
   Evaluate the tests and review areas where the children have done poorly. Review the pages and activities that stress the concepts tested.
   If necessary, when the children have reviewed sufficiently, administer the Alternate LIFEPAC Test. Follow the same procedures as used for the LIFEPAC Test.

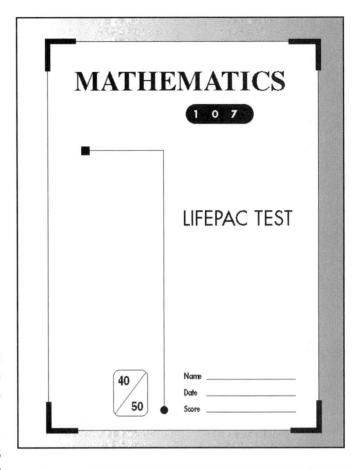

MATHEMATICS 107: LIFEPAC TEST

Put a circle around each number greater than 136.
(143) (162) 115 37 (192) 131

Put a circle around each number less than 159.
185 (104) (99) 160 194 (156)

Write 100's, 10's, 1's. (6 pts.)
107 = __100__ + __0__ + __7__
32 = _____ + __30__ + __2__

Write the number or number word.
fifteen __15__ 47 __forty-seven__
seventy-two __72__ 39 __thirty-nine__

Write the answer on the line ___.
   Mark spent 67¢ at the store. **Teacher check**
   He used __6__ dimes __1__ nickels __2__ pennies

Write the answer.
8+5=13 2+6+3=11 5+3+6=14 74+5=79 23+64=87

page 1 (one)

Write the answer.

12	11	10	12	11
− 6	− 8	− 3	− 9	− 4
6	3	7	3	7

Circle =, ≠.

7 + 3 (=, ≠) 9   Four plus three (=, ≠) eight.

12 − 6 (=, ≠) 5   Seven minus two (=, ≠) five.

Circle >, <.

185 (>, <) 176   3 + 5 (>, <) 2 + 7

191 (>, <) 200   6 − 3 (>, <) 5 − 4

Write the time.   **Write a.m. or p.m.**

Mother
makes our
dinner.

12:45   p.m.

Write the fraction.

$\frac{2}{3}$ $\frac{4}{5}$ = $\frac{2}{8}$ worms

= $\frac{3}{9}$ balls

page 2 (two)

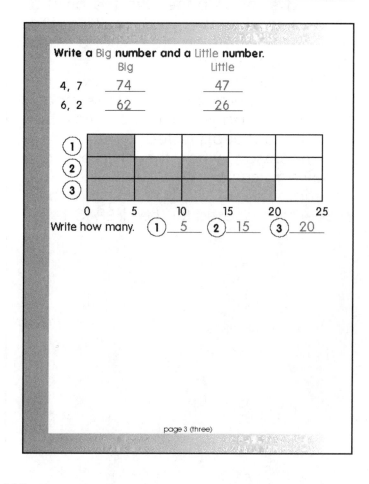

Write a Big number and a Little number.

	Big	Little
4, 7	74	47
6, 2	62	26

Write how many. ① 5 ② 15 ③ 20

page 3 (three)

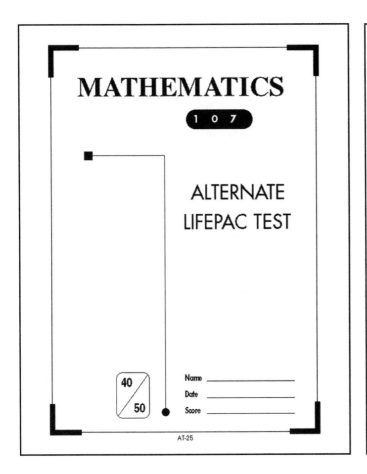

MATHEMATICS
1 0 7

ALTERNATE LIFEPAC TEST

40/50

Name _____
Date _____
Score _____

AT-25

MATHEMATICS 107: Alternate LIFEPAC TEST

Put a circle around each number greater than 148.

122   (172)   109   (150)   132   (198)

Put a circle around each number less than 159.

(116)   199   180   (152)   175   (130)

Write 100's, 10's, 1's. (6 pts.)

130 = __100__ + __30__ + __0__

46 = _____ + __40__ + __6__

Write the number or number word.

thirteen __13__   79 __seventy-nine__

eighty-six __86__   60 __sixty__

Write the answer on the line ___.

Mark spent 89¢ at the store.   Teacher check

He used __8__ dimes __1__ nickels __4__ pennies

Write the answer.

7	4	7	32	42
+ 3	+ 3	+ 1	+ 6	+ 36
10	+ 6	+ 8	38	78
	13	16		

AT-26

Write the answer.

10	12	11	11	12
− 5	− 4	− 3	− 6	− 7
5	8	8	5	5

Circle =, ≠.

5 + 9 (=, ≠) 15   Seven plus nine (=, ≠) seventeen.

14 − 6 (=, ≠) 8   Six minus zero (=, ≠) six.

Circle >, <.

163 (>, <) 158   6 + 2 (>, <) 5 + 4

194 (>, <) 200   9 − 4 (>, <) 11 − 9

Write the time.   Write a.m. or p.m.

We eat breakfast.

__8:15__   __a.m.__

Write the fraction.

2/4   4/6
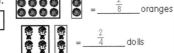
= 2/8 oranges
= 2/4 dolls

AT-27

Write a Big number and a Little number.

	Big	Little
7, 6	76	67
3, 1	31	13

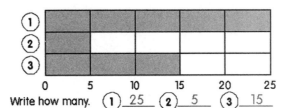

Write how many. (1) __25__ (2) __5__ (3) __15__

AT-28

Page 1: Fun With Numbers

CONCEPT(S): purpose of LIFEPAC, objectives

TEACHER GOAL(S): To teach the children
To know what is expected of the student in the LIFEPAC, and
To write first and last names correctly in manuscript.

MATERIALS/MANIPULATIVES:
pencils

TEACHING PAGE 1:

Turn to page 1. Point to the title and the memory verse and read them aloud. Allow time for the children to look through the LIFEPAC. Write the word *OBJECTIVES* on the board and have the children find the word on the page. Explain that the objectives tell the things the students will be expected to do in the LIFEPAC. Read each one and have the children repeat as they run their fingers under the sentence from left to right. Talk about the objectives so that the children will understand what they will be doing. Have each child write his name on the line.

FUN WITH NUMBERS

 My name is   Teacher check

 Memory Verse
"Let your light so shine before men, that they may see your good works, and glorify your Father which is in heaven."
Matthew 5:16

 **Objectives**

1. I can subtract to 18.
2. I can recognize zero as a place holder.
3. I can tell time to five minutes.
4. I can measure to the half-inch.
5. I can solve problems on the number line.

page 1 (one)

I. PART ONE

Page 2: Addition Facts

CONCEPT(S): addition facts to 18

TEACHER GOAL(S): To teach the children To review addition facts to 18 using number symbols and number words.

MATERIALS/MANIPULATIVES:
pencils, addition fact cards for 5's, 6's, 7's, 8's, and 9's

TEACHING PAGE 2:
Review the addition fact cards for 5's, 6's, 7's, 8's, and 9's with the students. Note for later review those facts that the students have not mastered.

Turn to page 2. Read the instructions with the students and have them read aloud the number word problems at the bottom of the page. Students may use their fact cards for the problems on this page. Review the fact cards again when the page is completed.

I. Part One

Write the answers.

9	8	5	9	8	9	7
+ 1	+ 2	+ 5	+ 3	+ 3	+ 4	+ 4
10	10	10	12	11	13	11

4	9	7	8	5	3	8
+ 7	+ 5	+ 3	+ 7	+ 9	+ 8	+ 8
11	14	10	15	14	11	16

5	6	9	7	4	1	7
+ 6	+ 8	+ 7	+ 7	+ 8	+ 9	+ 6
11	14	16	14	12	10	13

9	7	8	9	8	7	6
+ 6	+ 9	+ 5	+ 2	+ 4	+ 5	+ 6
15	16	13	11	12	12	12

Write the number words.

Five + six = ___eleven___ .	Nine + five = ___fourteen___ .
Three + two = ___five___ .	One + eight = ___nine___ .
Seven + eight = ___fifteen___ .	Four + five = ___nine___ .
Six + two = ___eight___ .	Eight + six = ___fourteen___ .

page 2 (two)

Page 3: Addition

CONCEPT(S): addition: columnar, two-digit

TEACHER GOAL(S): To teach the children To review addition of 3 one-digit numbers and 2 two-digit numbers, and To check answers by adding down and adding up.

MATERIALS/MANIPULATIVES:
pencils

TEACHING PAGE 3:

Turn to page 3. Read the directions at the top of the page with the students. Ask them how they can check to find out if their answers are right or wrong. (Add down and then add up.) Work the first problem in each row with the children and then let them complete the page independently. Monitor their work carefully to ensure that they are not simply copying the answers from adding down when they add up.

Add down. Add up.

12	10	16	14	12	14
6	3	7	2	5	9
2	5	2	4	4	0
+ 4	+ 2	+ 7	+ 8	+ 3	+ 5
12	10	16	14	12	14

11	12	13	12	14	7
3	4	8	2	5	4
3	1	5	7	3	2
+ 5	+ 7	+ 0	+ 3	+ 6	+ 1
11	12	13	12	14	7

42	68	75	99	89	77
40	61	53	79	38	73
+ 2	+ 7	+ 22	+ 20	+ 51	+ 4
42	68	75	99	89	77

90	95	94	96	86	103
60	93	74	15	31	63
+ 30	+ 2	+ 20	+ 81	+ 55	+ 40
90	95	94	96	86	103

page 3 (three)

Page 4: Subtraction, Number Order

CONCEPT(S): subtraction facts 13 to 15, ordinal numbers

TEACHER GOAL(S): To teach the children
To learn subtraction facts for 13, 14, and 15, and
To recognize ordinal (order) numbers as words.

MATERIALS/MANIPULATIVES:
pencils, subtraction fact cards 10's through 12's, new subtraction fact cards on 2 inch by 3 inch cardboard using facts as shown on page 4 - two sets may be made with and without the answer on the back of the card

TEACHING PAGE 4:
Review the subtraction fact cards *10's* through *12's* with the students. Note for later review those facts that the students have not mastered.

Turn to page 4. Review subtraction on the number line with the students. Have the children locate *13* on the number line. Then tell them to count to the left *3* places. Tell them to write the answer to *13* minus *3* (10) in the first box. Continue in this manner. Be sure the students are writing the correct answer for each problem. *Writing an incorrect answer will reinforce it in their minds.* Read the next set of directions. Tell the children to point to each balloon and say: "This is the first balloon, this is the second balloon..." and so on to the eleventh balloon. Then have them read each of the ordinal number words aloud. Read the directions again with the students and let them complete the exercise. Go over the subtraction fact cards for 10's through 12's with the students again.

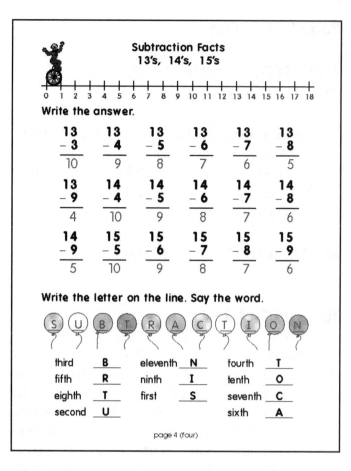

Subtraction Facts
13's, 14's, 15's

Write the answer.

13 −3 = 10	13 −4 = 9	13 −5 = 8	13 −6 = 7	13 −7 = 6	13 −8 = 5
13 −9 = 4	14 −4 = 10	14 −5 = 9	14 −6 = 8	14 −7 = 7	14 −8 = 6
14 −9 = 5	15 −5 = 10	15 −6 = 9	15 −7 = 8	15 −8 = 7	15 −9 = 6

Write the letter on the line. Say the word.

S U B T R A C T I O N

third	B	eleventh	N	fourth	T
fifth	R	ninth	I	tenth	O
eighth	T	first	S	seventh	C
second	U			sixth	A

page 4 (four)

Page 5: Subtraction

CONCEPT(S): subtraction facts 13 to 15

TEACHER GOAL(S): To teach the children
To practice subtraction facts for 13, 14, and 15.

MATERIALS/MANIPULATIVES:
pencils, subtraction fact cards 10 through 15, objects for counting

TEACHING PAGE 5:
Review fact cards for 10's through 15's. Use objects for counting to help students who are having difficulty with the concept(s).

Turn to page 5 and read the instructions with the students. Have them read the questions at the bottom of the page aloud. Students should be able to complete this page independently. Allow them to use their fact cards or objects for counting to complete the page.

Write the answer.

14 −6 (8)	14 −7 7	15 −9 (6)	14 −9 5	13 −4 9
13 −9 (4)	13 −5 (8)	15 −6 9	15 −7 (8)	13 −6 7
13 −3 (10)	13 −7 (6)	15 −5 (10)	14 −7 7	14 −4 (10)
15 −8 7	14 −5 9	14 −8 (6)	13 −9 (4)	13 −8 5

Count how many times the answer was 6.   3
Count how many times the answer was 8.   3
What answer was the smallest number?   4
What answer was the biggest number?   10
Circle all the even number answers green.

page 5 (five)

Page 6: Place Value

CONCEPT(S): place value for 1's, 10's, 100's

TEACHER GOAL(S): To teach the children To recognize the value of numbers in ones' place, tens' place, and hundreds' place.

MATERIALS/MANIPULATIVES:
pencils, paper, colored strips for counting that represent 1's, 10's, and 100's

TEACHING PAGE 6:
Write the number *132* on the board and have the students write it on paper. Ask the students to identify the number in the *hundreds'* place (1), *tens'* place (3), and *ones'* place (2). Write the number *146* on the board and have the students identify the *hundreds', tens',* and *ones'* places. Tell them to write the number *146* below the *123* on their papers so that the number places line up. Continue by dictating the numbers *23, 2, 157, 38, 4.* Have the students write the numbers one below the other so that the numbers line up by hundreds' place, tens' place and ones' place.

Turn to page 6 and direct the students' attention to the first problem. Ask the place value of the *3* and the *4.* Ask the children what the value is of the *3.* (The 3 is in the tens' place. The 3 means that there are 3 tens. Use the colored strips to count by 10 to 30. 3 tens are 30.) Have the students write the *30* in the tens' column. Ask them what the value is of the *4.* (The 4 is in the ones' place. The 4 means that there are 4 ones. Use the colored strips to count by 1 to 4. 4 ones is 4.) Have the students write the *4* in the *ones'* column. Continue in this manner using the counting strips to illustrate the problems.

Place Value

Write the 100's, 10's, 1's.

		100's		10's		1,s
34	=	_____	+	30	+	4
107	=	100	+	0	+	7
6	=	_____	+	_____	+	6
153	=	100	+	50	+	3
40	=	_____	+	40	+	0
179	=	100	+	70	+	9

Write the number.

56	=		+	50	+	6
149	=	100	+	40	+	9
102	=	100	+	0	+	2
113	=	100	+	10	+	3
20	=		+	20	+	0
163	=	100	+	60	+	3

We use ___0___ as a place holder.

page 6 (six)

Review the meaning of zero as a *place holder.* Emphasize that zero has no value but it holds a place. Discuss with the students the difference between 107 and 17, between 40 and 4. When this part of the exercise is complete read the other instructions on the page. Say to the children: "If there are 50 in the tens' column, how many ten are there?(5) What number do we write in the tens' place?(5) If there is a 6 in the ones' column, what number do we write in the ones' place (6)?" Have them complete this exercise and write the answer to the final problem.

SELF TEST 1:

CONCEPT(S): addition and subtraction facts, columnar and two-digit addition, place value

TEACHER GOAL(S): To teach the children To learn to check their progress periodically.

MATERIALS/MANIPULATIVES:
pencils

TEACHING PAGE 7:

Turn to page 7. Read the directions with the children. Be sure they understand what they are to do. You may repeat the directions but give no other help. Do not have the children check their own work. Check it as soon as you can and go over it with each child. Show him where he did well and where he needs extra help.

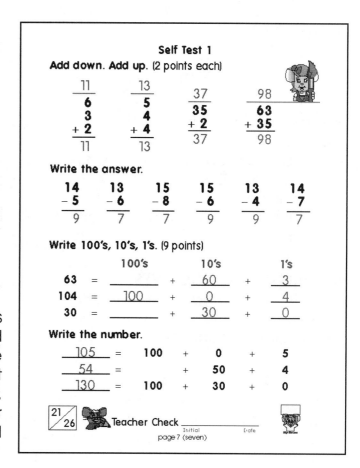

Self Test 1

Add down. Add up. (2 points each)

```
 11      13
  6       5      37       98
  3       4      35       63
 +2      +4      +2      +35
 ‾‾      ‾‾      ‾‾      ‾‾
 11      13      37       98
```

Write the answer.

```
 14      13      15      15      13      14
 -5      -6      -8      -6      -4      -7
 ‾‾      ‾‾      ‾‾      ‾‾      ‾‾      ‾‾
  9       7       7       9       9       7
```

Write 100's, 10's, 1's. (9 points)

		100's		10's		1's
63	=	_____	+	60	+	3
104	=	100	+	0	+	4
30	=	_____	+	30	+	0

Write the number.

105	=	100	+	0	+	5
54	=		+	50	+	4
130	=	100	+	30	+	0

$\frac{21}{26}$ Teacher Check _____

page 7 (seven)

115

II. PART TWO

Pages 8 and 9: Telling Time

CONCEPT(S): tell time to 5 minutes

TEACHER GOAL(S): To teach the children
To count by 5's,
To tell time to 5 minutes, and
To understand the meaning of a.m.
and p.m.

MATERIALS/MANIPULATIVES:
pencils, paper, student clock from
LIFEPAC 103 page 26 or any clock
suitable for student use

TEACHING PAGES 8 and 9:
Review the students' ability to tell time by having them move the hands on their clocks to represent different times to the hour (3:00, 5:00, 9:00, 12:00). Tell them that the little hand is called the *hour hand* because it tells the hour and the big hand is called the *minute hand* because it tells the minutes.

Turn to page 8. Call the students' attention to the clock at the top of the page and ask how many minutes are on a clock (60). Point out that the *1* on the clock also represents *5* minutes. Have the students count around the clock by *5's* saying "Five minutes, ten minutes, fifteen minutes" and so on to *60*. Using their own clocks, tell the students to put the hour (little) hand on the *3* and then, move the minute (big) hand around the clock while they count by *5's*. (five past three, ten past three, fifteen past three and so on to four o'clock) Go around the clock once more starting at four o'clock. Explain to the students that as the minute hand moves, the hour hand also moves but much more slowly. Have the children look at the first clock in the middle of the

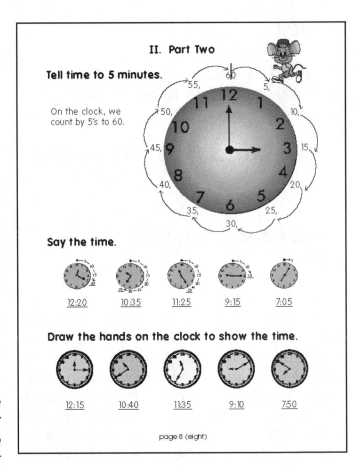

page. Show them that the hour hand is between the *12* and the *1*. Tell them that when we tell the time, we always say the hour that the hour (little) hand is moving *away from*. Have the children put their fingers on the *12* on the second clock and move their fingers clockwise around to the *10*. Have them read the time shown on the line. Ask the children to explain why it is 10:35 and not 11:35. Continue in this manner for the next three clocks. Read the next set of directions and have the children read the times below each clock. Have them fill in the hour hand first by moving their fingers around the clock in a clockwise motion until they find the correct spot for the hour hand. Have them follow the same clockwise direction to fill in the minute hand.

Turn to page 9. This page should be completed orally with the students. Have them move their fingers in a clockwise motion to show the correct time for the hour hand (the number the hand is moving *away from*). Tell them to write that number first. Next have the students count by *5's* to the position of the minute hand and write that number down. (Remind them to include the colon.) Finally, have the students say the time on the clock aloud (including a.m. or p.m.). Continue in this manner to write the times on all of the clocks. Review the meaning of a.m. and p.m. (before noon and after noon). Using their clocks, ask the children to illustrate midnight and noon on their clocks. Talk about how many hours on the clock before noon and how many hours on the clock after noon. Ask how many total hours in a day. Tell the children to look at the clocks on page 9 where they filled in the times. Tell them to write these times in the *before noon* or *after noon* columns at the bottom of the page.

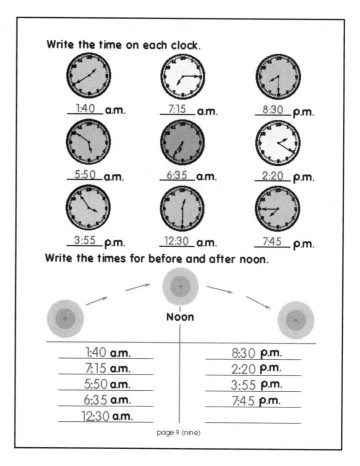

Pages 10 and 11: Calendar

CONCEPT(S): time on the calendar

TEACHER GOAL(S): To teach the children
To read and write time on a calendar.

MATERIALS/MANIPULATIVES:
pencils, current calendar showing
months of year, crayons

TEACHING PAGES 10 and 11:
Students will need to use a current calendar to complete this assignment. Use the calendar to review the name of the current year and the months of the year.

Turn to page 10 and read the title at the top of the page. Read the words *January* through *December* with the students and ask what they represent. Have the students fill in the name of today's month and today's year by copying from the current calendar. Read the next group of words with the students and ask what they represent. Tell the students to write the days of the week on the slanted lines above the calendar. Have the children continue answering questions and following instructions on pages 10 and 11 to complete the calendar on page 10. Let the children finish the exercise by drawing a picture of something fun that they did or plan to do in the current month.

Today's Month   Suggested answer

_____ MARCH _____ , _____ 200- _____
   (month)   (year)

Sunday	Monday	Tuesday	Wednesday	Thursday	Friday	Saturday
		1	2	3	4	5
6	7	8	9	10	11	12
13	14	15	(16)	17	18	19
20	21	22	23	24	25	26
27	28	29	30	31		

Write the name of today's month on the line.
January February March April May June July
August September October November December

Write the year.

Write the days of the week on the lines.
Sunday Monday Tuesday Wednesday
Thursday Friday Saturday

What is the name of the first day of this month?
_____ Tuesday _____

Put the number 1 on the calendar for the first day.

page 10 (ten)

How many days in this month? __31__

Write the numbers for the days on the calendar.
What is the name of the last day of this month?
_____ Friday _____

What is the date today? __March__ __16__ , __200-__
               month   day   year

Circle the date for today on the calendar.

How many more days in this month after today? __15__

Draw a picture of something fun that you did or are going to do this month. Tell what it is.

Teacher check

Something Fun _____

page 11 (eleven)

Page 12: Number Words

CONCEPT(S): reading and writing number words

TEACHER GOAL(S): To teach the children To read and write number words to ninety-nine.

MATERIALS/MANIPULATIVES: pencils, pencils

TEACHING PAGE 12:

Dictate the number words *one* through *nine* and have the students write them on paper. Check the spelling before proceeding.

Turn to page 12. Read the number words in the boxes with the children and then read the instructions. Point out the hyphen between the words and remind the students that the hyphens must be included for correct spelling. The students should be able to complete this page independently.

eleven	fourteen	seventeen
twelve	fifteen	eighteen
thirteen	sixteen	nineteen

Write the number on the line.

thirty-two	32	thirteen	13
fifteen	15	fifty-seven	57
eight	8	seventy	70
ninety-five	95	seventy-one	71
sixty	60	eighty-nine	89
forty-four	44	eleven	11

twenty	fifty	seventy
thirty	sixty	eighty
forty		ninety

Write the word on the line.

46	forty-six	84	eighty-four
3	three	26	twenty-six
27	twenty-seven	39	thirty-nine
91	ninety-one	40	forty
15	fifteen	7	seven
12	twelve	63	sixty-three

page 12 (twelve)

Page 13: Linear Measurement

CONCEPT(S): measuring inches

TEACHER GOAL(S): To teach the children To measure to the one-half inch.

MATERIALS/MANIPULATIVES:
pencils, cardboard, tracing paper, paste or glue, crayons

TEACHING PAGE 13:

Turn to page 13. Call the children's attention to the ruler at the bottom of the page. Have them count off one inch, two inches, and so on to six inches. Point to the first inch on the ruler. Ask the children how many parts the inch is divided into (2). Ask them if they remember how to write this as a fraction. (The number of parts is the bottom number of the fraction and the number of parts being talked about is the top number.) Point to the fraction and have them say *one-half inch*. Tell the children to point to the red lines on the ruler and ask what they represent. Ask them to count off one-half inch, one inch, one and one-half inches, two inches, and so on to six inches. Using the tracing paper, have the children copy the ruler on page 13 and paste it to a piece of cardboard. Read the rhyme at the top of the page. Tell the students to use their new rulers to draw straight lines going dot-to-dot. Have them use the rulers to measure the lines. Students who would like to may finish the page by naming and coloring the funny animal. Some students may want to use their rulers to make their own animals using the dot-to-dot method.

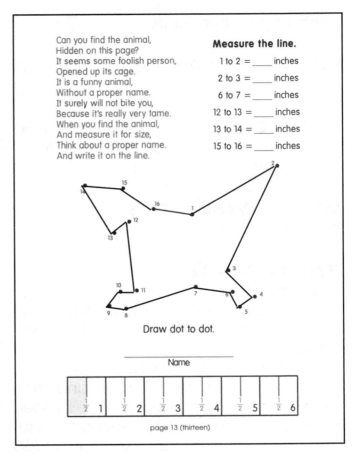

Can you find the animal,
Hidden on this page?
It seems some foolish person,
Opened up its cage.
It is a funny animal,
Without a proper name.
It surely will not bite you,
Because it's really very tame.
When you find the animal,
And measure it for size,
Think about a proper name.
And write it on the line.

Measure the line.

1 to 2 = _____ inches

2 to 3 = _____ inches

6 to 7 = _____ inches

12 to 13 = _____ inches

13 to 14 = _____ inches

15 to 16 = _____ inches

Draw dot to dot.

Name

page 13 (thirteen)

SELF TEST 2:

CONCEPT(S): time to 5 minutes, calendar, number words, measuring to one-half inch

TEACHER GOAL(S): To teach the children To learn to check their progress periodically.

MATERIALS/MANIPULATIVES:
pencils, current calendar

TEACHING PAGE 14:

Turn to page 14. Read the directions with the children. Be sure they understand what they are to do. You may repeat the directions but give no other help. Do not have the children check their own work. Check it as soon as you can and go over it with each child. Show him where he did well and where he needs extra help.

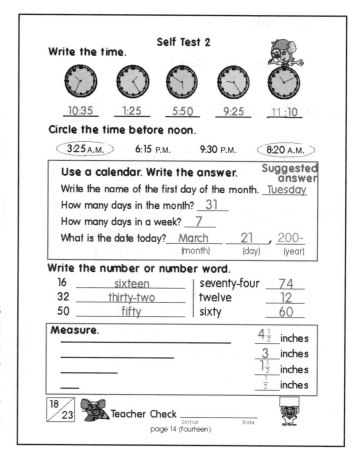

III. PART THREE

Page 15: Addition Facts

CONCEPT(S): addition facts to 18

TEACHER GOAL(S): To teach the children To review addition facts to 18.

MATERIALS/MANIPULATIVES: pencils, addition fact cards for 5's through 9's, objects for counting

TEACHING PAGE 15:

Turn to page 15. Dictate the following problems to the students and have them write them in the boxes. Remind the children to use the correct symbol for addition and to draw the line between the problem and the answer. (Seven plus three equals ten and so on.)

7	5	8	5	6
$+3$	$+4$	$+2$	$+9$	$+7$
10	9	10	14	13

Read the second set of directions and have the students complete the page independently. Students may use fact cards or objects for counting to complete the page, if necessary.

III. Part Three

Listen and write.

7	5	8	5	6
$+3$	$+4$	$+2$	$+9$	$+7$
10	9	10	14	13

Write the answer.

7	6	3	7	4	3	8
$+3$	$+6$	$+9$	$+7$	$+6$	$+2$	$+5$
10	12	12	14	10	5	13

7	8	6	9	6	0	5
$+4$	$+8$	$+0$	$+5$	$+2$	$+9$	$+6$
11	16	6	14	8	9	11

3	5	9	7	8	5	9
$+8$	$+3$	$+4$	$+6$	$+4$	$+5$	$+2$
11	8	13	13	12	10	11

4	9	8	3	9	5	6
$+3$	$+9$	$+7$	$+6$	$+5$	$+2$	$+9$
7	18	15	9	14	7	15

page 15 (fifteen)

Page 16: Subtraction Facts

CONCEPT(S): subtraction facts to 15

TEACHER GOAL(S): To teach the children To review subtraction facts to 15.

MATERIALS/MANIPULATIVES:
pencils, subtraction fact cards to 15's, objects for counting

TEACHING PAGE 16:
 Turn to page 16. Dictate the following problems to the students and have them write them in the boxes. Remind them to use the correct symbol for subtraction and to draw the line between the problem and the answer. (Twelve minus four equals eight and so on.)

12	9	13	15	11
− 4	− 6	− 5	− 8	− 7
8	3	8	7	4

Read the second set of directions and have the students complete the page independently. Students may use fact cards or objects for counting to complete the page, if necessary.

Listen and write.

12	9	13	15	11
− 4	− 6	− 5	− 8	− 7
8	3	8	7	4

Write the answer.

10	10	10	12	11	13	11
− 1	− 2	− 5	− 3	− 3	− 4	− 4
9	8	5	9	8	9	7

11	12	10	15	14	11	15
− 7	− 5	− 3	− 7	− 9	− 8	− 8
4	7	7	8	5	3	7

11	14	10	14	12	10	13
− 6	− 8	− 7	− 7	− 8	− 9	− 6
5	6	3	7	4	1	7

15	11	13	11	12	12	12
− 6	− 9	− 5	− 2	− 4	− 5	− 9
9	2	8	9	8	7	3

page 16 (sixteen)

Page 17: Subtraction Facts

CONCEPT(S): subtraction facts to 18

TEACHER GOAL(S): To teach the children To learn subtraction facts to 18.

MATERIALS/MANIPULATIVES:
pencils, new subtraction fact cards on 2 inch by 3 inch cardboard using the facts for 16's, 17's, and 18's shown on page 17, objects for counting

TEACHING PAGE 17:
Turn to page 17. Review subtraction on the number line with the students. Have the children complete the number facts using the number line. Have them write the missing numbers in the balloons. When the page is complete, review the new fact cards with the students. Have them use objects for counting to illustrate each fact card.

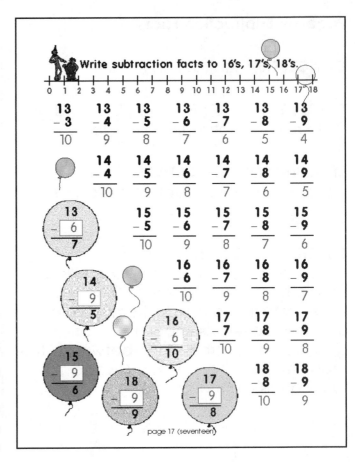

Page 18: Subtraction Facts

CONCEPT(S): subtraction facts to 18

TEACHER GOAL(S): To teach the children To review subtraction facts to 18.

MATERIALS/MANIPULATIVES:
pencils, subtraction fact cards to 18, objects for counting

TEACHING PAGE 18:

Review subtraction fact cards for 10's through 18's. Use objects for counting to help students who are having difficulty with the concept.

Turn to page 18 and read the instructions with the students. Have them read the questions at the bottom of the page aloud. Students should be able to complete this page independently. Allow them to use their fact cards or objects for counting to complete the page.

Write the answer.

10 − 8 2	12 − 5 ⑦	16 − 8 8	12 − 4 8	15 − 6 ⑨
7 − 7 0	4 − 1 ③	10 − 3 ⑦	6 − 4 2	11 − 2 ⑨
13 − 9 4	17 − 8 ⑨	16 − 7 ⑨	8 − 1 ⑦	15 − 7 8
9 − 5 4	14 − 9 ⑤	18 − 9 ⑨	14 − 7 ⑦	13 − 8 ⑤

Count how many times the answer was 9. __5__

Count how many times the answer was 7. __4__

What answer was the smallest number? __0__

What answer was the biggest number? __9__

Circle all the odd number answers purple.

page 18 (eighteen)

Page 19: Graphs and Number Lines

CONCEPT(S): locating numbers on a number line

TEACHER GOAL(S): To teach the children To locate numbers 0 to 20 on a number line.

MATERIALS/MANIPULATIVES:
pencils, addition and subtraction fact cards, number symbol cards for numbers 1 through 18 on 2 inch by 3 inch cardboard

TEACHING PAGE 19:

Turn to page 19 and read the title. Call the children's attention to the number line and ask the children to explain the pattern of the numbers (skip count by 5's). Ask what numbers are understood but not written on the line (1-4, 6-9, 11-14, 16-19). Tell the children to point to the approximate location of 3, 8, 14, 17 on the number line. Continue locating numbers until the children can find the locations easily. Read the directions on the page. Have the children follow the steps in the first example. Find the answer to the first fact (9) and point to its location on the number line. Write the numbers from the number line that are before and after 9 (5, 10). Circle the number that 9 is closest to (10). Tell the students to complete the page following the steps shown in the example.

When the page is complete, play a game of concentration with the students using selected addition or subtraction fact cards with matching number symbol cards.

Graphs and Number Lines

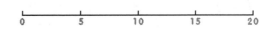

0   5   10   15   20

Add or subtract.
Find the answer on the number line.
Write the numbers from the number line that are before and after.
Circle the closest number.

6 + 3	8 − 4	7 + 1	6 + 6
5　9　(10)	0　4　(5)	5　8　(10)	(10)　12　15
5 + 8	9 − 2	5 − 3	9 + 9
10　13　(15)	(5)　7　10	(0)　2　5	15　18　(20)
7 + 4	4 + 9	8 + 9	8 − 2
(10)　11　15	10　13　(15)	(15)　17　20	(5)　6　10

page 19 (nineteen)

Page 20: Story Problems

CONCEPT(S): solving story problems

TEACHER GOAL(S): To teach the children
To apply new skills to story problems.

MATERIALS/MANIPULATIVES:

pencils, crayons, pitcher and glasses, box of 13 cookies, pennies, ruler, four squares of paper - each $1\frac{1}{2}$ inch square (students should make these squares), a piece of paper 8 inches wide, clock, objects for counting

TEACHING PAGE 20:

Turn to page 20. The students should read the story problems carefully. Students may work together or the teacher may help in the reading. Students will need to use manipulatives to make the exercises more interesting and meaningful to them. Each exercise should be expressed as a written problem to be complete. For example: The students may begin the first problem by filling a pitcher with six glasses of water. They may then pour two glasses, one for Jane and one for Lisa. They may count what is left by pouring the remaining water into glasses until the pitcher is empty. This should then be thought out: *Lisa started with 6 glasses; she subtracted 2 glasses; she has 4 glasses left* or *6 – 2 = 4.* Continue to use manipulatives and illustrations to solve each of the story problems.

Read. Write the problem. Circle the answer.

Lisa had a pitcher that had six glasses of water in it. She poured one glass for Jane and one glass for herself. How many glasses of water were left in the pitcher?

$$\begin{array}{r} 6 \\ -\ 2 \\ \hline 4 \end{array} \text{ glasses}$$

Jerry had a box of thirteen cookies. He ate six of the cookies. How many were left in the box?

$$\begin{array}{r} 13 \\ -\ 6 \\ \hline 7 \end{array} \text{ cookies}$$

Lucy and Marie wanted to buy new pencils. The pencils cost 65 cents. Lucy had 32 cents. Marie had 36 cents. Did they have enough money to buy the pencils?

$$\begin{array}{r} 32 \\ +\ 36 \\ \hline 68 \end{array} ¢$$

yes

Ben wanted to draw four squares on his paper. Each square was one and one-half inch. The paper was eight inches wide. Could Ben draw four squares across the paper?

yes

Jack's dad left for work at 6:00 A.M. He came home at 6:00 P.M. How many hours from the time Jack's dad left to the time he came home?

12 hours

Richard and John were standing in line in front of the school door. John was eighth in line and Richard was second in line. Who would go through the door sooner, Richard or John?

Richard

page 20 (twenty)

SELF TEST 3:

CONCEPT(S): addition and subtraction facts, locating numbers on the number line, solving story problems.

TEACHER GOAL(S): To teach the children To learn to check their progress periodically.

MATERIALS/MANIPULATIVES: pencils, objects for counting

TEACHING PAGE 21:

Turn to page 21. Read the directions with the children. Be sure they understand what they are to do. You may repeat the directions but give no other help. Do not have the children check their own work. Check it as soon as you can and go over it with each child. Show him where he did well and where he needs extra help. Students may use objects for counting to complete the addition and subtraction facts. Dictate the following problems for *Listen and write*.

$$
\begin{array}{cccc}
4 & 16 & 3 & 10 \\
+\,5 & -\,8 & +\,9 & -\,4 \\
\hline
9 & 8 & 12 & 6
\end{array}
$$

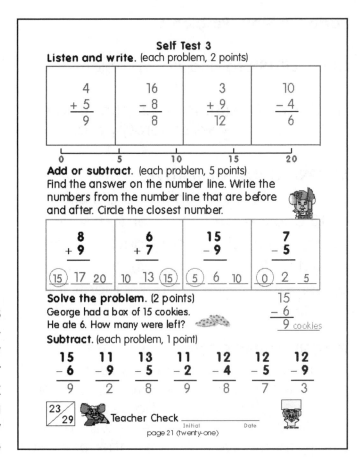

IV. PART FOUR

Pages 22 and 23: Fractions

CONCEPT(S): fractions to fifths

TEACHER GOAL(S): To teach the children To divide a whole into parts and to express the part as a fraction.

MATERIALS/MANIPULATIVES:
pencils, paper for writing, scissors, ruler, several pieces of paper approximately 4 inches square (construction paper is suggested)

TEACHING PAGES 22 and 23:

Have each student take a square of paper and hold it up. Ask the children how many pieces of paper they each have (one whole). Tell them to use pencil and ruler to draw a line dividing the square into two equal parts. Then have them cut the paper along the line they have drawn. Ask the children to describe the two parts that they now have (one-half and one-half). Tell them to write the fraction $\frac{1}{2}$ on the paper. Have them hold up the two pieces and point to the bottom number of the fraction (2); have them hold up one piece and point to the top number of the fraction (1). Tell the children to put the two parts together again and ask what they have ($\frac{2}{2}$ or one whole). Continue this exercise with the squares dividing them into thirds, fourths, and fifths. Have the students write the fractions for $\frac{1}{3}$, $\frac{1}{4}$, $\frac{1}{5}$ and then put the parts back together for one whole.

Turn to pages 22 and 23. Point to the illustration at the top of page 22 and tell

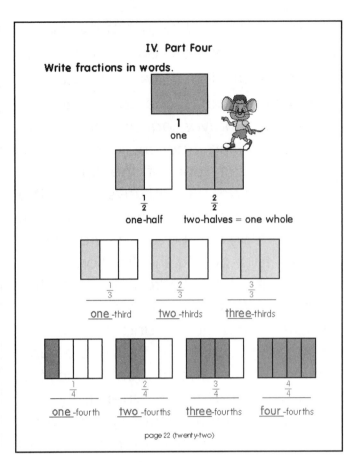

129

the children to read the number symbol *1* and the number word *one*. Point to the second illustration. Have the students identify and read: $\frac{1}{2}$, one-half, $\frac{2}{2}$, two *halves, two halves equal one whole.* Continue writing fractions and number words for each one of the figures on pages 22 and 23. Always have the students write the bottom number (denominator) first. Be sure they are using hyphens to separate the fraction words. Encourage the children to use the squares that they cut out to further illustrate each figure and to cut out additional figures for the circle and triangle. Have the students read the fractions aloud to be sure they are pronouncing them correctly. *Students should begin to realize that the bottom number(denominator) of a fraction sounds like the ordinal (order) number and the top number (numerator) sounds like the cardinal (counting) number.* Discuss the singular and plural forms of the words at the top of page 23.

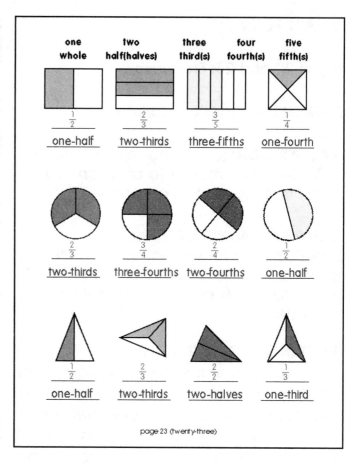

Page 24: Count to 200

CONCEPT(S): counting to 200

TEACHER GOAL(S): To teach the children
  To review counting to 200

MATERIALS/MANIPULATIVES:
pencils, chart of numbers to 200 -
LIFEPAC 107 page 2

TEACHING PAGE 24:
  Turn to page 24. Read the directions and have the students complete the page independently. They may use the chart of numbers if necessary. When the page is complete have the children count aloud from 100 to 200.

Count to 200. Write the number in each box.

100	101	102	103	104	105	106	107	108	109
110	111	112	113	114	115	116	117	118	119
120	121	122	123	124	125	126	127	128	129
130	131	132	133	134	135	136	137	138	139
140	141	142	143	144	145	146	147	148	149
150	151	152	153	154	155	156	157	158	159
160	161	162	163	164	165	166	167	168	169
170	171	172	173	174	175	176	177	178	179
180	181	182	183	184	185	186	187	188	189
190	191	192	193	194	195	196	197	198	199
200									

Write the missing numbers.

128 , _129_ , _130_ , 131 , 132 , _133_ , 134

153 , _154_ , 155 , _156_ , 157 , _158_ , 159

170 , _171_ , _172_ , 173 , 174 , _175_ , _176_

184 , _185_ , 186 , _187_ , 188 , _189_ , 190

194 , 195 , _196_ , _197_ , 198 , _199_ , _200_

page 24 (twenty-four)

Page 25: Shapes

CONCEPT(S): flat and solid shapes, ordinal numbers

TEACHER GOAL(S): To teach the children
To review flat and solid shapes, and
To review ordinal (order) numbers.

MATERIALS/MANIPULATIVES:
pencils, blue and red crayons

TEACHING PAGE 25:

Turn to page 25 and read the title and rhyme with the students. Ask them what the train is pulling and to identify each object (shapes - cylinder, triangle, sphere, square, cone, cube, circle, rectangle).

Review the difference between flat and solid shapes and have the students identify each shape as flat or solid. Read the directions and the ordinal number words aloud with the students. Go down the list of questions to ensure that each student can read the name of each shape. Have the students complete the page but monitor their work to be sure they understand the directions.

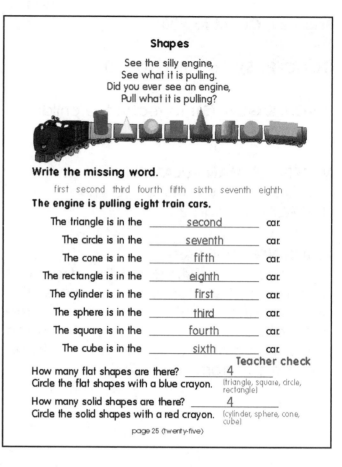

Shapes

See the silly engine,
See what it is pulling.
Did you ever see an engine,
Pull what it is pulling?

Write the missing word.

first second third fourth fifth sixth seventh eighth

The engine is pulling eight train cars.

The triangle is in the _____second_____ car.
The circle is in the _____seventh_____ car.
The cone is in the _____fifth_____ car.
The rectangle is in the _____eighth_____ car.
The cylinder is in the _____first_____ car.
The sphere is in the _____third_____ car.
The square is in the _____fourth_____ car.
The cube is in the _____sixth_____ car.

Teacher check

How many flat shapes are there? _____4_____
Circle the flat shapes with a blue crayon. (triangle, square, circle, rectangle)

How many solid shapes are there? _____4_____
Circle the solid shapes with a red crayon. (cylinder, sphere, cone, cube)

page 25 (twenty-five)

Page 26: Facts, Number Words, Flat Shapes

CONCEPT(S): facts, number words, flat shapes

TEACHER GOAL(S): To teach the children
To solve word problems in addition and subtraction, and
To identify flat shapes and colors.

MATERIALS/MANIPULATIVES:
pencils, paper, yellow, blue, green, and orange crayons, addition and subtraction fact cards

TEACHING PAGE 26:

Review the number words on page 12 with the students. Let them use this page as a reference for the assignment on page 26.

Turn to page 26. Read the directions with the students and ask them to identify the shapes on the page. Have the children read several of the word problems aloud. Be sure they understand that the answers must be written as words. Some students will be able to write the answers immediately. Other students may need to convert the number word problems to number symbol problems on separate paper and then write the number word answer on page 26. *Three plus four equals_____. 3 + 4 = 7 Three plus four equals seven.* When the number word problems are solved, have the students complete the page by coloring the flat shapes as instructed. Spend some time reviewing addition and subtraction fact cards.

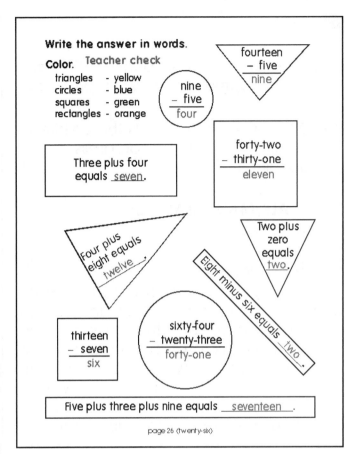

SELF TEST 4:

CONCEPT(S): fractions, count to 200, flat and solid shapes, number word problems

TEACHER GOAL(S): To teach the children To learn to check their progress periodically.

MATERIALS/MANIPULATIVES:
pencils, paper

TEACHING PAGE 27:

Turn to page 27. Read the directions with the children. Be sure they understand what they are to do. Students may write the word problems at the bottom of the page as number symbol problems on another piece of paper. They may then write the correct number word on the test. You may repeat the directions but give no other help. Do not have the children check their own work. Check it as soon as you can and go over it with each child. Show him where he did well and where he needs extra help.

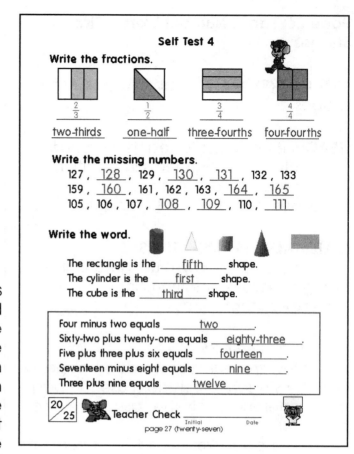

134

V. PART FIVE

Page 28: Numbers to 200

CONCEPT(S): number value to 200

TEACHER GOAL(S): To teach the children To identify numbers greater than and less than, before and after, and missing numbers to 200.

MATERIALS/MANIPULATIVES: pencils, chart of numbers to 200, counting strips for 100's, 10's, 1's

TEACHING PAGE 28:

Review the chart of numbers with the students. Have them identify numbers in the hundreds', tens', and ones' places. Ask them why *172* is greater than *156* (number in tens' place is larger). If they cannot give a good explanation, have them illustrate *172* and *156* using the strips.

Turn to page 28 and read each set of directions with the students. Be sure they understand what they are to do. The children should not use the chart of numbers unless the teacher is using it with them for instruction purposes.

V. Part Five

Put a circle around each number greater than 150.
128 (155) 150 110 (182) (176)

Put a circle around each number less than 150.
(117) (145) 150 159 162 (139)

Put a circle around the number
that comes after 144.   that comes after 175.
143 (145)   (176) 174
that comes after 188.   that comes after 158.
(189) 187   (159) 157

Put a circle around the number
that comes before 129. that comes before 190.
(128) 130   (189) 191
that comes before 124. that comes before 161.
(123) 125   162 (160)

Write the missing numbers.
127 , 128 , _129_ , _130_ , _131_ , 132 , 133
185 , _186_ , 187 , _188_ , 189 , _190_ , 191
190 , _191_ , _192_ , 193 , _194_ , _195_ , 196
176 , _177_ , 178 , 179 , _180_ , _181_ , _182_
_166_ , 167 , _168_ , 169 , _170_ , _171_ , 172

page 28 (twenty-eight)

Page 29: Operation Symbols

CONCEPT(S): operations symbols (+, –, =, ≠, >, <)

TEACHER GOAL(S): To teach the children To write the correct operation symbols for number symbol and number word problems.

MATERIALS/MANIPULATIVES:
pencils, paper

TEACHING PAGE 29:
Turn to page 29. Talk to the children about the meaning of operation symbols and review the operation symbols that they have learned thus far. Spend some time on the signs for greater than (>) and less than (<). Point out to the students that the open side is always toward the larger number.

Dictate the following to the children and have them write the numbers on paper: *twelve is greater than six (12 > 6), five is greater than three (5 > 3), two is less than four (2 < 4), seven is less than thirteen (7 < 13).* Continue working with the children until they are using the symbols (>, <) correctly. Read the directions for each section on page 29 and have the students complete the page.

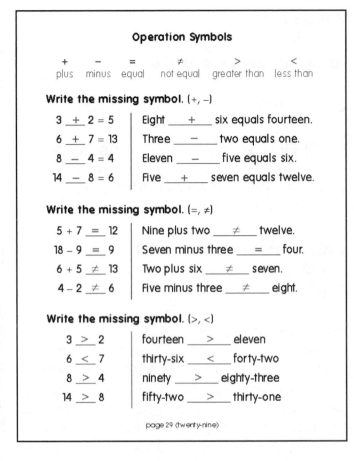

Operation Symbols

+	–	=	≠	>	<
plus	minus	equal	not equal	greater than	less than

Write the missing symbol. (+, –)

3 <u>+</u> 2 = 5 | Eight <u>+</u> six equals fourteen.
6 <u>+</u> 7 = 13 | Three <u>–</u> two equals one.
8 <u>–</u> 4 = 4 | Eleven <u>–</u> five equals six.
14 <u>–</u> 8 = 6 | Five <u>+</u> seven equals twelve.

Write the missing symbol. (=, ≠)

5 + 7 <u>=</u> 12 | Nine plus two <u>≠</u> twelve.
18 – 9 <u>=</u> 9 | Seven minus three <u>=</u> four.
6 + 5 <u>≠</u> 13 | Two plus six <u>≠</u> seven.
4 – 2 <u>≠</u> 6 | Five minus three <u>≠</u> eight.

Write the missing symbol. (>, <)

3 <u>></u> 2 | fourteen <u>></u> eleven
6 <u><</u> 7 | thirty-six <u><</u> forty-two
8 <u>></u> 4 | ninety <u>></u> eighty-three
14 <u>></u> 8 | fifty-two <u>></u> thirty-one

page 29 (twenty-nine)

Page 30: Estimation and Skip Counting

CONCEPT(S): estimation, skip counting

TEACHER GOAL(S): To teach the children
To estimate groups of objects, and
To learn to count quickly by grouping
numbers by 2's, 5's, and 10's.

MATERIALS/MANIPULATIVES:
pencils, an assortment of objects for
counting - pennies, beans, buttons, bottle
caps, wrapper twists

TEACHING PAGE 30:
Place a group of objects (about
twenty) in front of the students. Have
them count the set by 1's, taking one
object at a time (1-2-3 and so on). Ask
them if they think there might be an
easier way to count the set. Talk to them
about skip counting. Have the children
count the set again by 2's, taking two
objects at a time (2-4-6 and so on). Ask
which method they think is easier and
quicker.

Turn to page 30. Read the title, the
instructions, and the words at the top of
each column with the students. Complete the page in the following manner.
Select a set of objects and have the
children write the name in the first column
(buttons). Ask them how they would like
to skip count this set (by 2's, by 5's, by
10's) and write that in the next column.
Have the students estimate the number of
objects in the set and write the number
under *estimate*. Next, have them count
(by 2's, 5's, or 10's) and write the count
under *how many*. Finally, have them circle
the correct operation symbol (>,<). Work
with the students as they complete the
skip counting exercise. Use a variety of
objects and numbers of objects to make
the exercise more interesting. Have the
students review addition by completing

the problems at the bot-tom of the page.
*If the count is 21 and the student counts
by 5's, the student would count 5, 10, 15,
20 plus 1 more (21).*

Skip Counting

Count how many.   Teacher check

Name	By 2's, 5's, 10's	Estimate	Circle	How Many?
buttons	5's	25	(>) <	21
			> <	
			> <	
			> <	
			> <	
			> <	
			> <	
			> <	

Write the answer.

2 4 + 3 9	3 5 + 2 10	8 1 + 4 13	26 + 31 57	43 + 60 103	13 + 26 39	84 + 13 97

page 30 (thirty)

137

Page 31: Number Order, Number Words, Facts

CONCEPT(S): number order, number words, addition and subtraction facts

TEACHER GOAL(S): To teach the children
To understand number order,
To solve problems using number words, and
To practice addition and subtraction facts.

MATERIALS/MANIPULATIVES:
pencils, number symbol cards for 0 through 9

TEACHING PAGE 31:
Place the number symbol cards in front of the students. Give the students sets of two cards at random. Have them use the cards to make the larger number and the smaller number. Example: *1* and *5* can be *15* or *51*. Discuss the use of *0*. Example: *0* and *3* can be *03* (0 has no value) or *30* (0 holds the ones' place).

Turn to page 31. Read the instructions on the page with the students. Monitor the students work for the first exercise and then let them complete the page independently.

Write the largest number.

0, 5	8, 1	1, 3	4, 3
50	81	31	43

Write the numbers in order from small to large.

136	45	92	11	131	143
11	45	92	131	136	143

Write the number word.

Six plus seven equals _____thirteen_____.

Twelve minus five equals _____seven_____.

Eight plus four equals _____twelve_____.

Nine minus three equals _____six_____.

Write the answer.

$\begin{array}{r} 6 \\ +3 \\ \hline 9 \end{array}$	$\begin{array}{r} 9 \\ +5 \\ \hline 14 \end{array}$	$\begin{array}{r} 2 \\ +4 \\ \hline 6 \end{array}$	$\begin{array}{r} 8 \\ +1 \\ \hline 9 \end{array}$	$\begin{array}{r} 7 \\ +3 \\ \hline 10 \end{array}$	$\begin{array}{r} 2 \\ +6 \\ \hline 8 \end{array}$	$\begin{array}{r} 9 \\ +3 \\ \hline 12 \end{array}$
$\begin{array}{r} 7 \\ -2 \\ \hline 5 \end{array}$	$\begin{array}{r} 3 \\ -1 \\ \hline 2 \end{array}$	$\begin{array}{r} 18 \\ -9 \\ \hline 9 \end{array}$	$\begin{array}{r} 14 \\ -7 \\ \hline 7 \end{array}$	$\begin{array}{r} 11 \\ -3 \\ \hline 8 \end{array}$	$\begin{array}{r} 7 \\ -4 \\ \hline 3 \end{array}$	$\begin{array}{r} 12 \\ -5 \\ \hline 7 \end{array}$

page 31 (thirty-one)

Page 32: Place Value

CONCEPT(S): place value for ones, tens, and hundreds

TEACHER GOAL(S): To teach the children To understand place value for ones, tens, and hundreds.

MATERIALS/MANIPULATIVES: pencils, counting strips for 1's, 10's, 100's

TEACHING PAGE 32:
Turn to page 32. Review the counting strips with the children. Have the students use the counting strips to illustrate each number in the exercises. Begin with *162*. Have the students select one *100* strip, six *10's* strips and two *1's* strips. Ask them the value of the *100* strip (100) and have them write that down, the value of the six *10's* strips (60) and write that down, the value of the two *1's* strips (2) and write that down. Use the reverse method for the second exercise. Have them select a *100* strip, four *10's* strips, and six *1's* strips. Ask them to count the number and write it on the blank.

Place Value

Write the 100's, 10's, 1's.

		100's		10's		1,s
162	=	100	+	60	+	2
45	=		+	40	+	5
6	=		+		+	6
70	=		+	70	+	0
103	=	100	+	0	+	3
66	=		+	60	+	6

Write the number.

146	=	100	+	40	+	6
37	=		+	30	+	7
8	=		+		+	8
109	=	100	+	0	+	9
50	=		+	50	+	0
195	=	100	+	90	+	5

page 32 (thirty-two)

Page 33: Patterns

CONCEPT(S): concepts and patterns

TEACHER GOAL(S): To teach the children
To read for understanding,
To find a pattern, and
To tell what comes next in the pattern.

MATERIALS/MANIPULATIVES:
pencils, objects that show a pattern such as little to big, crayons

TEACHING PAGE 33:
Place a group of objects in some type of pattern (little to big) in front of the students and ask them to identify the pattern. Ask them to describe what they think would be the next object in the pattern. Do the same thing again using another pattern (colors of blue, green, blue, green) and ask the children to describe what they think would be the next object in this pattern. Continue with a number of patterns until the children are successful in describing the next object.

Turn to page 33. Read the directions at the top of the page and tell the children to work as many of the problems as they can. Students should be encouraged to think independently, to find the pattern on their own and to draw or write the next object. Allow the students to complete the page alone whether or not errors are made. When they are finished, review the work and help them find the patterns where they had difficulty.

Find the pattern. Draw or write what comes next.

22 ,	24 ,	26 ,	28
$\frac{1}{5}$	$\frac{2}{5}$	$\frac{3}{5}$	$\frac{4}{5}$
$\begin{array}{r} 5 \\ +1 \\ \hline 6 \end{array}$	$\begin{array}{r} 5 \\ +2 \\ \hline 7 \end{array}$	$\begin{array}{r} 5 \\ +3 \\ \hline 8 \end{array}$	$\begin{array}{r} 5 \\ +4 \\ \hline 9 \end{array}$
4 < 5	5 < 6	6 < 7	7 < 8
three,	five,	seven,	nine
$\begin{array}{r} 13 \\ -5 \\ \hline 8 \end{array}$	$\begin{array}{r} 13 \\ -6 \\ \hline 7 \end{array}$	$\begin{array}{r} 13 \\ -7 \\ \hline 6 \end{array}$	$\begin{array}{r} 13 \\ -8 \\ \hline 5 \end{array}$
Tuesday,	Wednesday,	Thursday,	Friday

page 33 (thirty-three)

SELF TEST 5:

CONCEPT(S): number order, operation symbols, place value, patterns, facts

TEACHER GOAL(S): To teach the children To learn to check their progress periodically.

MATERIALS/MANIPULATIVES: pencils

TEACHING PAGE 34:

Turn to page 34. Read the directions with the children. Be sure they understand what they are to do. You may repeat the directions but give no other help. Do not have the children check their own work. Check it as soon as you can and go over it with each child. Show him where he did well and where he needs extra help.

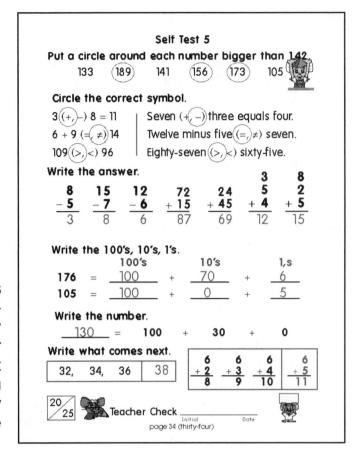

LIFEPAC TEST AND ALTERNATE TEST 108

CONCEPT(S): addition, subtraction, place value, time, number words, fractions, listening skills

TEACHER GOALS: To teach the children
To learn to check their own progress periodically.

MATERIALS/MANIPULATIVES:
pencils, rulers

TEACHING the LIFEPAC TEST:

Administer the test in at least two sessions.

Read all of the directions on each page as the children prepare to do it. Be sure that they understand what they are being asked to do. Give no help except with directions. Go over each page with the child as soon as possible after you check it so that he can see where he did well and where he needs more work.

Evaluate the tests and review areas where the children have done poorly. Review the pages and activities that stress the concepts tested.

If necessary, when the children have reviewed sufficiently, administer the Alternate LIFEPAC test. Follow the same procedures as used for the LIFEPAC Test.

LIFEPAC Test
Listen and write. (page 3 - first)
There are four boxes. Draw a circle, a square, a rectangle, and a triangle in that order. Circle the shape in the third box.
Listen and write. (page 3 - second)
Write these problems.

$$\begin{array}{c} 9 \\ +3 \\ \hline 12 \end{array} \qquad \begin{array}{c} 11 \\ -7 \\ \hline 4 \end{array} \qquad \begin{array}{c} 29 > 19 \\ 2 + 6 \neq 10 - 3 \end{array}$$

Four plus eight equals twelve.

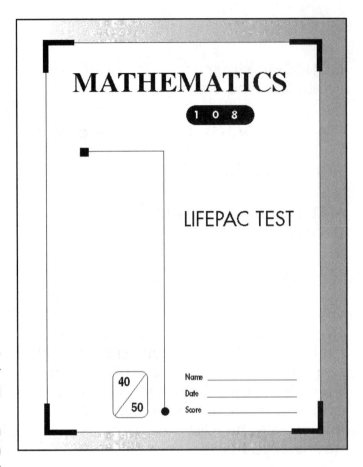

MATHEMATICS 108: LIFEPAC TEST

Add down. Add up. (5 points)

$$\begin{array}{c} 12 \\ \hline 3 \\ 5 \\ +4 \\ \hline 12 \end{array} \quad \begin{array}{c} 9 \\ \hline 2 \\ 4 \\ +3 \\ \hline 9 \end{array} \quad \begin{array}{c} 59 \\ 56 \\ +3 \\ \hline 59 \end{array} \quad \begin{array}{c} 97 \\ 35 \\ +62 \\ \hline 97 \end{array} \quad \begin{array}{c} 89 \\ 51 \\ +38 \\ \hline 89 \end{array}$$

Subtract.

$$\begin{array}{c} 15 \\ -8 \\ \hline 7 \end{array} \quad \begin{array}{c} 17 \\ -9 \\ \hline 8 \end{array} \quad \begin{array}{c} 13 \\ -6 \\ \hline 7 \end{array} \quad \begin{array}{c} 18 \\ -9 \\ \hline 9 \end{array} \quad \begin{array}{c} 16 \\ -8 \\ \hline 8 \end{array}$$

Write 100's, 10's, 1's. (each row 1 point)

		100's		10's		1's
47	=	_____	+	40	+	7
105	=	100	+	0	+	5
63	=	_____	+	60	+	3

Write the number.

180	=	100	+	80	+	0
94	=		+	90	+	4

page 1 (one)

Alternate LIFEPAC Test
Listen and write. (page 3 - first)
There are four boxes. Draw a square, a triangle, a rectangle, and a circle in that order. Circle the shape in the second box.
Listen and write. (page 3 - second)
Write these problems.

$$\begin{array}{r} 7 \\ +6 \\ \hline 13 \end{array} \qquad \begin{array}{r} 15 \\ -9 \\ \hline 6 \end{array} \qquad \begin{array}{l} 52 < 63 \\ 5 + 5 \neq 12 - 4 \end{array}$$

Thirteen minus five equals eight.

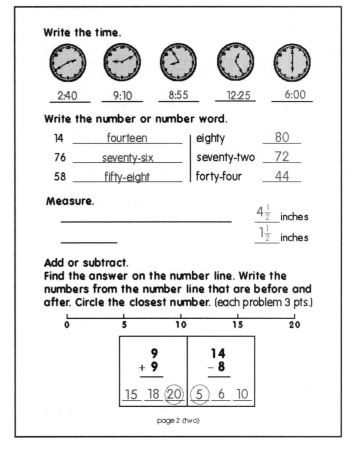

Write the time.

2:40   9:10   8:55   12:25   6:00

Write the number or number word.

14	fourteen	eighty	80
76	seventy-six	seventy-two	72
58	fifty-eight	forty-four	44

Measure.

_____ $4\frac{1}{2}$ inches

_____ $1\frac{1}{2}$ inches

Add or subtract.
Find the answer on the number line. Write the numbers from the number line that are before and after. Circle the closest number. (each problem 3 pts.)

0   5   10   15   20

$$\begin{array}{r} 9 \\ +9 \end{array} \qquad \begin{array}{r} 14 \\ -8 \end{array}$$

15   18   (20)   (5)   6   10

page 2 (two)

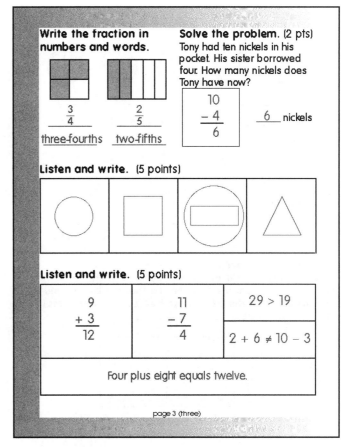

Write the fraction in numbers and words.

$\frac{3}{4}$ $\frac{2}{5}$

three-fourths   two-fifths

Solve the problem. (2 pts)
Tony had ten nickels in his pocket. His sister borrowed four. How many nickels does Tony have now?

$$\begin{array}{r} 10 \\ -4 \\ \hline 6 \end{array}$$

6 nickels

Listen and write. (5 points)

Listen and write. (5 points)

| $\begin{array}{r} 9 \\ +3 \\ \hline 12 \end{array}$ | $\begin{array}{r} 11 \\ -7 \\ \hline 4 \end{array}$ | $29 > 19$ |
| | | $2 + 6 \neq 10 - 3$ |

Four plus eight equals twelve.

page 3 (three)

MATHEMATICS
1 0 8

ALTERNATE
LIFEPAC TEST

40
50

Name _____

Date _____

Score _____

AT-29

MATHEMATICS 108: Alternate LIFEPAC TEST

Add down. Add up. (5 points)

12	14			
4	**5**	48	66	87
2	**1**	**46**	**41**	**63**
+ 6	**+ 8**	**+ 2**	**+ 25**	**+ 24**
12	14	48	66	87

Subtract.

18	**16**	**14**	**17**	**16**
− 9	**− 7**	**− 8**	**− 9**	**− 7**
9	9	6	8	9

Write 100's, 10's, 1's. (each row 1 point)

	100's	10's	1's
150 =	100 +	50 +	0
65 =	____ +	60 +	5
29 =	____ +	20 +	9

Write the number.

43 = ____ + **40** + **3**

175 = **100** + **70** + **5**

AT-30

Write the time.

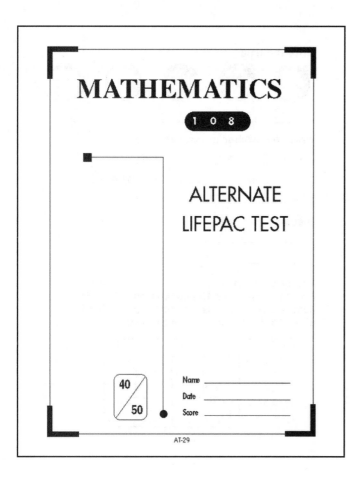

1:40   10:55   6:05   12:25   3:00

Write the number or number word.

35	thirty-five	forty-one	41
13	thirteen	sixty-three	63
38	thirty-eight	fifty	50

Measure.

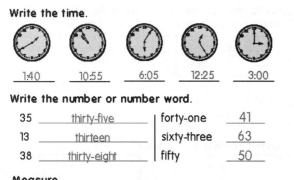

$3\frac{1}{2}$ inches

$\frac{1}{2}$ inches

Add or subtract.
Find the answer on the number line. Write the numbers from the number line that are before and after. Circle the closest number. (each problem 3 pts.)

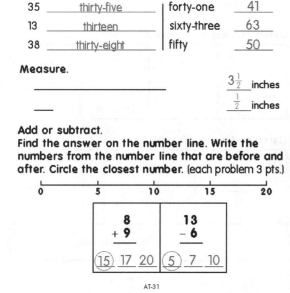

8	**13**
+ 9	**− 6**
(15) 17 20	(5) 7 10

AT-31

Write the fraction.

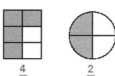

$\frac{4}{6}$ $\frac{2}{4}$

four-sixths   two-fourths

Solve the problem. (2pts)
Tony had eight nickels in his pocket. His sister borrowed five. How many nickels does Tony have now?

8
− 5
3

____3____ nickels

Listen and write. (5 points.)

Listen and write. (5 points.)

7	15	52 < 63
+ 6	− 9	
13	6	5 + 5 ≠ 12 − 4

Thirteen minus five equals eight.

AT-32

Page 1: Fun With Numbers

CONCEPT(S): purpose of LIFEPAC, objectives

TEACHER GOAL(S): To teach the children
To know what is expected of the student in the LIFEPAC, and
To write first and last names correctly in manuscript.

MATERIALS/MANIPULATIVES:
pencils

TEACHING PAGE 1:

Turn to page 1. Point to the title and the memory verse and read them aloud. Allow time for the children to look through the LIFEPAC. Write the word *OBJECTIVES* on the board and have the children find the word on the page. Explain that the objectives tell the things the students will be expected to do in the LIFEPAC. Read each one and have the children repeat as they run their fingers under the sentence from left to right. Talk about the objectives so that the children will understand what they will be doing. Have each child write his name on the line.

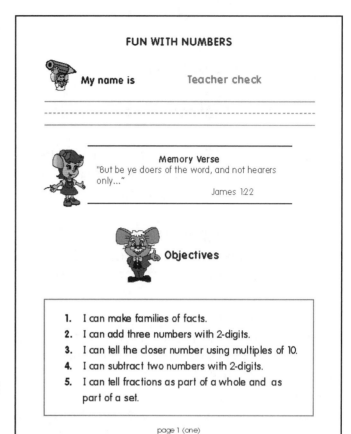

FUN WITH NUMBERS

My name is   Teacher check

Memory Verse
"But be ye doers of the word, and not hearers only..."
James 1:22

Objectives

1. I can make families of facts.
2. I can add three numbers with 2-digits.
3. I can tell the closer number using multiples of 10.
4. I can subtract two numbers with 2-digits.
5. I can tell fractions as part of a whole and as part of a set.

page 1 (one)

I. PART ONE

Page 2: Families of Facts

CONCEPT(S): fact families for addition

TEACHER GOAL(S): To teach the children To understand the relationship between certain addition facts.

MATERIALS/MANIPULATIVES:
pencils, number symbol cards for 0 through 9, paper, objects for counting.

TEACHING PAGE 2:
Talk to the children about the meaning of families and how families are related to each other. Tell them that certain facts are related and that we call them *families of facts*. Select two number symbol cards (2, 4). Place the number cards in front of the students with the *4* below the *2* so that it resembles a number fact. Tell the children to look at the cards and to write an addition number fact on paper using the two number symbols. Ask them for the answer (6). Now place the *2* below the *4* and tell the children to write another addition fact. Ask if the answer is the same (6). Point out to them that this is a family of facts because no matter which way we place the number symbols *2* and *4*, when we add the answer is always *6*.

Turn to page 2. Have the children complete the first row of fact families by illustrating them using objects for counting. Have them say each one aloud. "One plus two equals three. Two plus one equals three." Continue to work each row of problems with the students having them fill in the missing numbers in the boxes.

I. Part One

Complete the family of facts for addition.

1	2	4	3	5	8	7	4
+2	+1	+3	+4	+8	+5	+4	+7
3	3	7	7	13	13	11	11
2	5	6	3	4	2	8	6
+5	+2	+3	+6	+2	+4	+6	+8
7	7	9	9	6	6	14	14
9	3	7	2	3	5	5	2
+3	+9	+2	+7	+5	+3	+2	+5
12	12	9	9	8	8	7	7
4	6	3	0	7	1	6	5
+6	+4	+0	+3	+1	+7	+5	+6
10	10	3	3	8	8	11	11

page 2 (two)

Page 3: Families of Facts

CONCEPT(S): fact families for subtraction

TEACHER GOAL(S): To teach the children To understand the relationship between certain subtraction facts.

MATERIALS/MANIPULATIVES:
pencils, number symbol cards for 0 through 9, paper, objects for counting.

TEACHING PAGE 3:
Review with the children that certain facts are related and that we call them *families of facts*. On page 2, the children learned about families for addition. There are also families for subtraction. Select two number symbol cards (7, 4). Place the number cards in front of the students with the *4* below the *7* so that it resembles a number fact. Ask the children to look at the cards and to write a subtraction number fact on paper using the two number symbols. Ask them for the answer (3) and then give them the number symbol card for *3*. Now place the *3* below the *7* and tell the children to write another subtraction fact. Ask what the answer is this time (4). Point out to them that this is a family of facts because if we subtract *3* from *7*, the answer is *4* and if we subtract *4* from *7*, the answer is *3*.

Turn to page 3. Have the children complete the first row of fact families by illustrating them using objects for counting. Have them say each one aloud. "Six minus two equals four. Six minus four equals two." Continue to work each row of problems with the students having them fill in the missing numbers in the boxes.

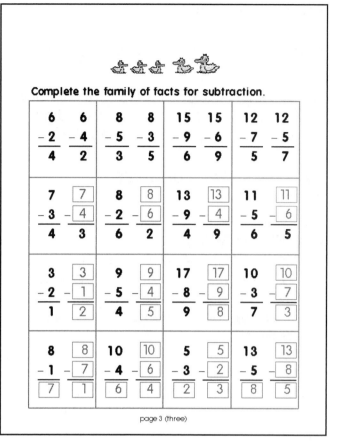

page 3 (three)

147

Page 4: Addition

CONCEPT(S): addition review

TEACHER GOAL(S): To teach the children To review addition facts, two-digit addition, and columnar addition.

MATERIALS/MANIPULATIVES:
pencils, crayons

TEACHING PAGE 4:
Turn to page 4. Tell the children that this is a page of addition problems and review with them how to do each problem. Have them locate the clown and the balloons.

Tell them that the answer to each problem is on one of the balloons. As they complete each problem, they should find the balloon that has the answer and color the balloon the same color as the square in the problem.

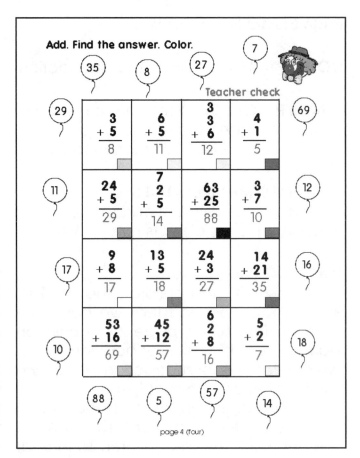

Page 5: Addition

CONCEPT(S): two digit columnar addition

TEACHER GOAL(S): To teach the children To add two-digit numbers in columns.

MATERIALS/MANIPULATIVES: pencils

TEACHING PAGE 5:

Turn to page 5 and read the title with the students. Point to the first illustration and read the directions aloud. Ask the children to read the numbers in the first illustration and then to say the numbers in the ones' place and the tens' place. Explain to the children that the red lines are for their subtotals. In the second and third illustrations, show the students that the answer has a number in the hundreds' place.

Have the children complete the page using the same steps as shown in the illustration. Monitor their work closely. Be sure that they always begin with the ones' column and that they put the subtotal number on the red lines.

Three Number Addition

Add ones' place first. Add tens' place second.

5 24 9	
35	
+ 40	
99	

4 33 3	
10	
+ 72	
115	

8 56 7	
31	
+ 42	
129	

7   32   3   8   63   7   5   15   7
41   24   42
+ 25   + 70   + 61
98   157   118

9   71   3   11   64   4   8   31   9
22   50   58
+ 35   + 23   + 70
128   137   159

3   15   6   5   22   7   5   26   7
21   35   31
+ 43   + 10   + 52
79   67   109

page 5 (five)

SELF TEST 1:

CONCEPT(S): families of facts, addition

TEACHER GOAL(S): To teach the children
To learn to check their progress
periodically.

MATERIALS/MANIPULATIVES:
pencils

TEACHING PAGE 6:

Turn to page 6. Read the directions with the children. Be sure they understand what they are to do. You may repeat the directions but give no other help. Do not have the children check their own work. Check it as soon as you can and go over it with each child. Show him where he did well and where he needs extra help.

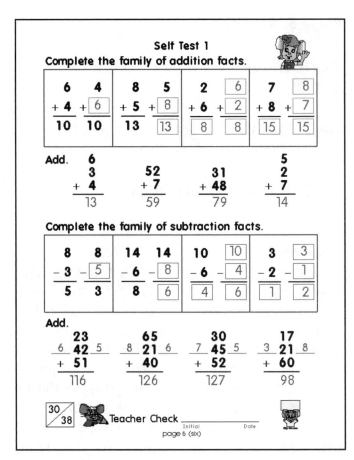

II. PART TWO

Page 7: Counting, Number Words

CONCEPT(S): count to 200, number words to ninety-nine

TEACHER GOAL(S): To teach the children
To write number symbols to 199, and
To write number words to ninety-nine.

MATERIALS/MANIPULATIVES:
pencils

TEACHING PAGE 7:
Turn to page 7. Read the directions to each section with the students. In the second section, have them copy the number words on each line. Remind them to use good posture and to hold their pencils correctly.

II. Part Two

Write the missing numbers.

0	1	2	3	4	5	6	7	8	9
10	11	12	13	14	15	16	17	18	19

90	91	92	93	94	95	96	97	98	99
100	101	102	103	104	105	106	107	108	109

Write the number words.

20 twenty
30 thirty
40 forty
50 fifty
60 sixty
70 seventy
80 eighty
90 ninety

page 7 (seven)

Page 8: Skip Counting and Shapes

CONCEPT(S): even and odd numbers; skip counting by 2's, 5's, and 10's; shapes

TEACHER GOAL(S): To teach the children
To identify even and odd numbers;
To count by 2's, 5's, and 10's; and
To recognize shapes.

MATERIALS/MANIPULATIVES:
pencils, crayons

TEACHING PAGE 8:
Turn to page 8. Tell the children to follow the numbers around the outside of the page with their fingers, counting aloud as they point. Read each direction in the first box with the children, allowing them time to complete the exercise before reading the next direction. Be sure they understand the direction before they begin. Read the direction to the second box. Have the children identify the shapes by name when they have completed the exercise.

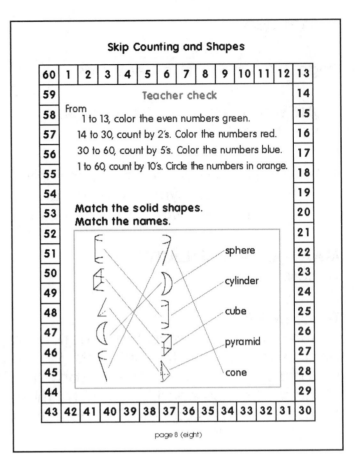

page 8 (eight)

152

Page 9: Counting Money

CONCEPT(S): dimes, nickels, pennies

TEACHER GOAL(S): To teach the children To group money into dimes, nickels, and pennies.

MATERIALS/MANIPULATIVES:
pencils, pennies, nickels, dimes (play money)

TEACHING PAGE 9:
    Review counting money. Ask the children to show how many pennies in a nickel, how many pennies in a dime, and how many nickels in a dime.
    Turn to page 9. Read the instructions aloud with the children. Point to the ball and ask the children the price shown for the ball. Tell them to use the play money to show how many dimes, nickels, and pennies they would need to pay for the ball. Have them write the numbers on the lines. (2 dimes, 1 nickel, 2 pennies *or* 2 dimes, 7 pennies) Once the children understand what they are to do, have them complete the page independently. Allow them to answer in their own way. There are several possible answers to each amount. You may review the different possibilities once the students have completed the page.

How Many?	Dimes	Nickels	Pennies
		Suggested answers	
27¢	2	+ 1	+ 2
40¢	2	+ 3	+ 5
68¢	4	+ 4	+ 8
72¢	6	+ 2	+ 2
59¢	3	+ 4	+ 9
87¢	8	+ 1	+ 2
13¢		+ 2	+ 3
8¢		+	+ 8

page 9 (nine)

Page 10: Telling Time

CONCEPT(S): calendars, clocks

TEACHER GOAL(S): To teach the children
To read a calendar, and
To tell time to 5 minutes.

MATERIALS/MANIPULATIVES:
pencils, current calendar, classroom clock

TEACHING PAGE 10:
Review a current calendar with the children. Show them how the months appear in order and how the days of the week are in order.

Turn to page 10. Tell the children to use the current calendar to put the months and the days of the week in order from *1* to *12* and *1* to *7*. Ask them to write the current date and day of the week where indicated. Read the last direction and have the students write the time on the lines. Then tell them to look at the classroom clock and to draw the hour hand and the minute hand to show the current time to the nearest five minutes.

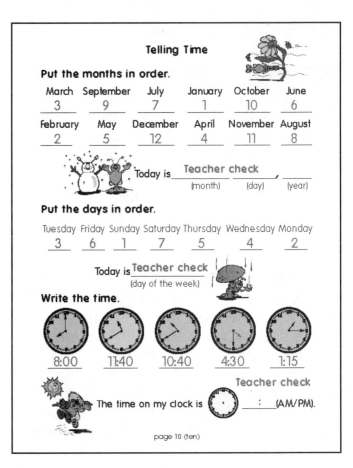

Page 11: Number Words

CONCEPT(S): number words to ninety-nine

TEACHER GOAL(S): To teach the children To write number words to ninety-nine.

MATERIALS/MANIPULATIVES:
pencils

TEACHING PAGE 11:
Turn to page 11. Read the directions with the students. In the first section, have them write the numbers in words on each line. Remind the students to use hyphens. Tell them to use good posture and to hold their pencils correctly. In the last section, the children should write the numbers using number symbols and then arrange the numbers in number order.

Write the number words.

11 eleven 15 fifteen
12 twelve 16 sixteen
13 thirteen 17 seventeen
14 fourteen 18 eighteen
19 nineteen
24 twenty-four
52 fifty-two

Write the number symbols on the line.
Write the numbers in order from smallest to largest.

thirteen 13 thirty-five 35 eleven 11 fifteen 15
forty-three 43 eighty-nine 89 seven 7 sixty-two 62

7 11 13 15 35 43 62 89

page 11 (eleven)

Page 12: Families of Facts

CONCEPT(S): families of facts for addition and subtraction

TEACHER GOAL(S): To teach the children To recognize families of facts for addition and subtraction.

MATERIALS/MANIPULATIVES:
pencils, paper bag, 2 sets of ten cardboard squares approximately 2 inches by 2 inches in size with the numbers 0 through 9 written on them

TEACHING PAGE 12:

Turn to page 12. Ask the children if they remember solving problems using families of facts (pages 2 and 3). Explain to them that addition and subtraction can also make a family. Put the twenty cardboard squares in the bag and mix them well. Ask the students to reach in the bag and to select any two numbers. (Example: 5, 3) Tell the children that they can write a family of four facts using these two numbers. Have them begin by writing the two addition facts in the first box on page 12. (They will need to write four facts in each box so be sure they do not write too large.) (5 + 3 = 8, 3 + 5 = 8) Now ask them to write two subtraction facts using the numbers 3, 5, and 8. (*Do not pull an 8 from the bag.*) (8 − 5 = 3, 8 − 3 = 5) Point out to the children that these facts are related because we have used the same three numbers (3, 5, 8) in each one. Tell the students to draw two more numbers and to write four facts in the next box. As the children complete each box, have them arrange the numbers drawn in number order. When all the numbers have been drawn, all the boxes on page 12 should be filled and there should be two sets of numbers from *0* through *9*.

Family of Facts
Suggested answer

$\begin{array}{r}3\\+5\\\hline8\end{array}$ $\begin{array}{r}5\\+3\\\hline8\end{array}$	$\begin{array}{r}7\\+2\\\hline9\end{array}$ $\begin{array}{r}2\\+7\\\hline9\end{array}$	$\begin{array}{r}6\\+6\\\hline12\end{array}$ $\begin{array}{r}6\\+6\\\hline12\end{array}$	$\begin{array}{r}8\\+0\\\hline8\end{array}$ $\begin{array}{r}0\\+8\\\hline8\end{array}$
$\begin{array}{r}8\\-3\\\hline5\end{array}$ $\begin{array}{r}8\\-5\\\hline3\end{array}$	$\begin{array}{r}9\\-2\\\hline7\end{array}$ $\begin{array}{r}9\\-7\\\hline2\end{array}$	$\begin{array}{r}12\\-6\\\hline6\end{array}$ $\begin{array}{r}12\\-6\\\hline6\end{array}$	$\begin{array}{r}8\\-0\\\hline8\end{array}$ $\begin{array}{r}8\\-8\\\hline0\end{array}$
$\begin{array}{r}4\\+3\\\hline7\end{array}$ $\begin{array}{r}3\\+4\\\hline7\end{array}$	$\begin{array}{r}9\\+5\\\hline14\end{array}$ $\begin{array}{r}5\\+9\\\hline14\end{array}$	$\begin{array}{r}1\\+2\\\hline3\end{array}$ $\begin{array}{r}2\\+1\\\hline3\end{array}$	
$\begin{array}{r}7\\-4\\\hline3\end{array}$ $\begin{array}{r}7\\-3\\\hline4\end{array}$	$\begin{array}{r}14\\-5\\\hline9\end{array}$ $\begin{array}{r}14\\-9\\\hline5\end{array}$	$\begin{array}{r}3\\-1\\\hline2\end{array}$ $\begin{array}{r}3\\-2\\\hline1\end{array}$	
$\begin{array}{r}9\\+8\\\hline17\end{array}$ $\begin{array}{r}8\\+9\\\hline17\end{array}$	$\begin{array}{r}7\\+0\\\hline7\end{array}$ $\begin{array}{r}0\\+7\\\hline7\end{array}$	$\begin{array}{r}1\\+4\\\hline5\end{array}$ $\begin{array}{r}4\\+1\\\hline5\end{array}$	
$\begin{array}{r}17\\-9\\\hline8\end{array}$ $\begin{array}{r}17\\-8\\\hline9\end{array}$	$\begin{array}{r}7\\-0\\\hline7\end{array}$ $\begin{array}{r}7\\-7\\\hline0\end{array}$	$\begin{array}{r}5\\-1\\\hline4\end{array}$ $\begin{array}{r}5\\-4\\\hline1\end{array}$	

page 12 (twelve)

SELF TEST 2:

CONCEPT(S): skip counting, number words, counting money, telling time

TEACHER GOAL(S): To teach the children To learn to check their progress periodically.

MATERIALS/MANIPULATIVES: pencils

TEACHING PAGE 13:

Turn to page 13. Read the directions with the children. Be sure they understand what they are to do. You may repeat the directions but give no other help. Do not have the children check their own work. Check it as soon as you can and go over it with each child. Show him where he did well and where he needs extra help.

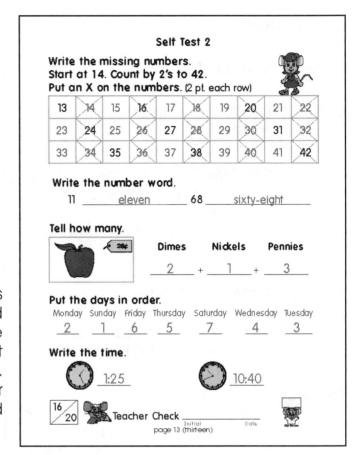

Self Test 2

Write the missing numbers.
Start at 14. Count by 2's to 42.
Put an X on the numbers. (2 pt. each row)

13	14	15	16	17	18	19	20	21	22
23	24	25	26	27	28	29	30	31	32
33	34	35	36	37	38	39	40	41	42

Write the number word.
11 ____eleven____   68 ____sixty-eight____

Tell how many.

Dimes   Nickels   Pennies
__2__ + __1__ + __3__

Put the days in order.
Monday Sunday Friday Thursday Saturday Wednesday Tuesday
__2__   __1__   __6__   __5__   __7__   __4__   __3__

Write the time.
1:25   10:40

16/20 Teacher Check _____
                     Initial   Date
page 13 (thirteen)

III. PART THREE

Page 14: Number Order

CONCEPT(S): before and after, greater than and less than

TEACHER GOAL(S): To teach the children
To write numbers that come before and after, and
To identify numbers that are greater than and less than.

MATERIALS/MANIPULATIVES:
pencils; chart of numbers from LIFEPAC 107, page 3 (or any chart that shows number order from 1 to 200)

TEACHING PAGE 14:
Turn to page 14. Ask the children to look at the first example. Ask them to explain how they know that *128* is before *129*. Have them point to the numbers in the hundreds' place and tens' place and show that they are the same. Have them point to the number in the ones' place and say, "Eight comes before nine." Children who are able may complete the page independently. If children are having difficulty, have them go through the steps of comparing number place from hundreds to tens to ones. Students may use the chart of numbers if necessary.

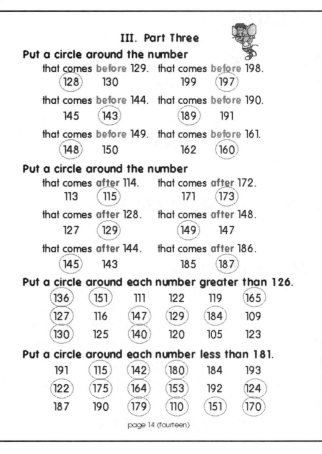

III. Part Three

Put a circle around the number

that comes before 129. that comes before 198.
(128) 130   199   (197)

that comes before 144. that comes before 190.
145   (143)   (189) 191

that comes before 149. that comes before 161.
(148) 150   162   (160)

Put a circle around the number

that comes after 114. that comes after 172.
113   (115)   171   (173)

that comes after 128. that comes after 148.
127   (129)   (149) 147

that comes after 144. that comes after 186.
(145) 143   185   (187)

Put a circle around each number greater than 126.
(136) (151) 111 122 119 (165)
(127) 116 (147) (129) (184) 109
(130) 125 (140) 120 105 123

Put a circle around each number less than 181.
191 (115) (142) (180) 184 193
(122) (175) (164) (153) 192 (124)
187 190 (179) (110) (151) (170)

page 14 (fourteen)

Page 15: Number Order

CONCEPT(S): number order using a scale of zero to ten

TEACHER GOAL(S): To teach the children To identify number value in relation to zero and ten.

MATERIALS/MANIPULATIVES: pencils

TEACHING PAGE 15:

Turn to page 15 and tell the students to look at the number line at the top of the page. Ask them to read the first question and to locate *4* on the number line. Tell them to write the answer *0* or *10* on the blank. Let them continue to answer the questions to the first section in the same manner. Read the next two directions. Have the students read the number order words aloud and then complete the page.

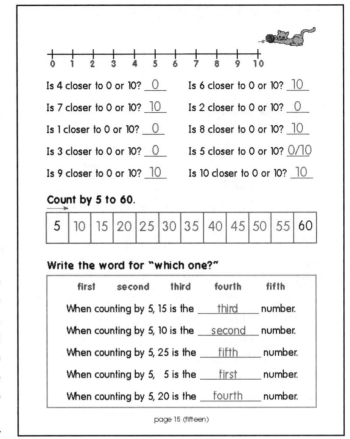

Is 4 closer to 0 or 10? __0__   Is 6 closer to 0 or 10? __10__

Is 7 closer to 0 or 10? __10__   Is 2 closer to 0 or 10? __0__

Is 1 closer to 0 or 10? __0__   Is 8 closer to 0 or 10? __10__

Is 3 closer to 0 or 10? __0__   Is 5 closer to 0 or 10? __0/10__

Is 9 closer to 0 or 10? __10__   Is 10 closer to 0 or 10? __10__

Count by 5 to 60.

5	10	15	20	25	30	35	40	45	50	55	60

Write the word for "which one?"

first	second	third	fourth	fifth

When counting by 5, 15 is the ___third___ number.

When counting by 5, 10 is the ___second___ number.

When counting by 5, 25 is the ___fifth___ number.

When counting by 5, 5 is the ___first___ number.

When counting by 5, 20 is the ___fourth___ number.

page 15 (fifteen)

159

Pages 16 and 17: Data and Graphs

CONCEPT(S): recording data, making a graph

TEACHER GOAL(S): To teach the children
To record data on paper, and
To illustrate the data on a graph.

MATERIALS/MANIPULATIVES:
pencils, 18 building blocks, crayons

TEACHING PAGES 16 and 17:

Turn to page 16 and read the title at the top of the page. Have the children read the first column of words (first through tenth) and the words at the top of the other three columns. Give the students the eighteen blocks and tell them that they will have ten chances to build a tower with their blocks using all eighteen blocks. When the tower falls over they must begin again. Tell them that they will record how many blocks tall the tower was each time and how many blocks were left. They should then write a number fact using the numbers from each try. (For example: First try, used 8 blocks. There were 10 left. 18 - 8 = 10) Allow the students to continue until the tenth try is completed and the data is recorded.

Turn to page 17. Read the first instruction and have the students write *first* through *tenth* on the blanks at the left side of the graph. Have the children compare the number line at the top of the page to the numbers at the bottom of the graph. Ask them how many blocks they used for their first tower. (Example: 8) Have them locate *8* (approximately) on the graph after the word *first*. Tell them to draw a line where *8* is located and then to color in after *first* to the line they drew. Continue doing this until the graph is completed coloring in the number for "how many blocks tall" for each try, *first* through *tenth*.

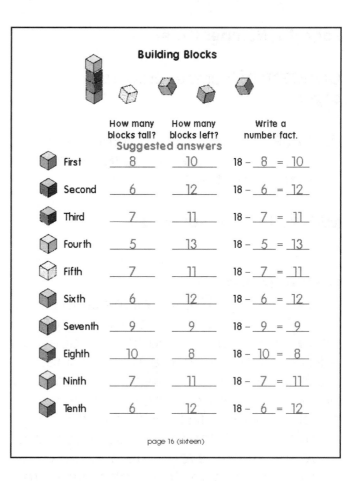

Building Blocks

	How many blocks tall?	How many blocks left?	Write a number fact.
	Suggested answers		
First	8	10	18 − 8 = 10
Second	6	12	18 − 6 = 12
Third	7	11	18 − 7 = 11
Fourth	5	13	18 − 5 = 13
Fifth	7	11	18 − 7 = 11
Sixth	6	12	18 − 6 = 12
Seventh	9	9	18 − 9 = 9
Eighth	10	8	18 − 10 = 8
Ninth	7	11	18 − 7 = 11
Tenth	6	12	18 − 6 = 12

page 16 (sixteen)

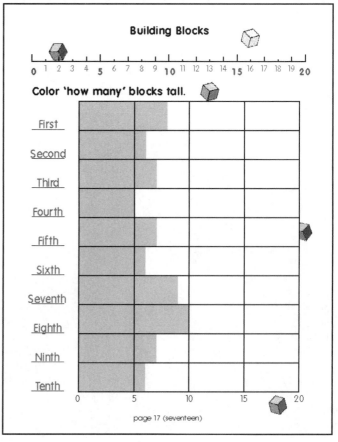

page 17 (seventeen)

Page 18: Addition Facts

CONCEPT(S): addition facts to 18

TEACHER GOAL(S): To teach the children
To review addition facts for 10
through 18.

MATERIALS/MANIPULATIVES:
pencils, addition fact cards

TEACHING PAGE 18:
Turn to page 18. Read the directions and have the students complete the page. Students may use the answer key to check their own answers. Those students who have not mastered their facts should spend some additional time at the end of the lesson reviewing their fact cards.

Write + or – in the box.

1 $+$ 2 = 3   4 $+$ 5 = 9   5 $+$ 3 = 8   9 $-$ 6 = 3

8 $-$ 7 = 1   9 $-$ 9 = 0   4 $-$ 1 = 3   6 $+$ 3 = 9

1 $+$ 6 = 7   2 $-$ 1 = 1   3 $+$ 2 = 5   2 $+$ 2 = 4

2 $+$ 4 = 6   4 $+$ 4 = 8   4 $-$ 1 = 3   7 $-$ 3 = 4

Find the sum. Write the numbers.

7 +5	6 +4	9 +2	5 +5	3 +8	9 +3	8 +2
12	10	11	10	11	12	10
6 +5	7 +3	8 +4	6 +6	5 +8	7 +6	8 +6
11	10	12	12	13	13	14
6 +8	9 +4	8 +8	9 +5	4 +9	9 +9	7 +8
14	13	16	14	13	18	15
6 +7	9 +8	8 +5	9 +6	8 +7	8 +3	7 +9
13	17	13	15	15	11	16

page 18 (eighteen)

Page 19: Story Problems

CONCEPT(S): solving story problems

TEACHER GOAL(S): To teach the children
To apply math skills to story problems, and
To read for comprehension.

MATERIALS/MANIPULATIVES:
pencils; dimes, nickels, pennies; ruler from LIFEPAC 108, page 13 (or any 6 inch ruler)

TEACHING PAGE 19:
Turn to page 19. The first group of problems require the students to convert dimes, nickels, and pennies to cents before finding the answer. Read the problems with the students and have them follow the correct steps to find the solutions. Students should use rulers to draw the lines. Read the final set of directions. Have the students write a story problem in words and write a number fact.

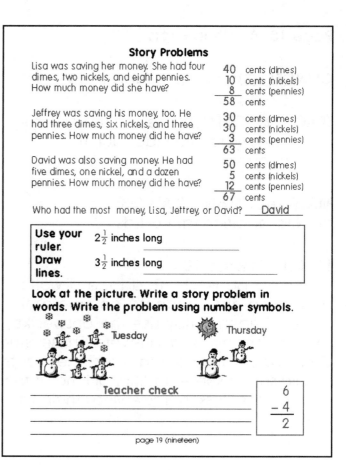

Page 20: Mathematics Symbols

CONCEPT(S): symbols (+, –, =, ≠, >, <)

TEACHER GOAL(S): To teach the children To read and write operation symbols for plus and minus, equal and not equal, greater than and less than.

MATERIALS/MANIPULATIVES:
pencils

TEACHING PAGE 20:
Turn to page 20. Tell the children that you will dictate six problems to them. They should write the problems in number symbols horizontally (across) in the boxes.
Remind them to use the correct symbols to write each problem.
Dictate: (Six plus eight equals fourteen.)

6 + 8 = 14	11 – 7 = 4
7 – 2 = 5	4 + 8 = 12
4 + 6 = 10	16 – 9 = 7

Review the symbols (equal) =, (not equal) ≠, (greater than) >, and (less than) <. Complete the first problem in each section with the students. Those who are able may complete the page independently. Continue working with the other students to help them understand the concept.

Listen and write.

6 + 8 = 14	11 – 7 = 4
7 – 2 = 5	4 + 8 = 12
4 + 6 = 10	16 – 9 = 7

Circle =, ≠, >, <.

3 + 4 (**=**, ≠) 7	5 + 4 (>, **<**) 10
9 + 5 (=, **≠**) 12	6 + 2 (**>**, <) 7
13 – 7 (=, **≠**) 6	9 – 6 (**>**, <) 2
15 – 7 (=, **≠**) 9	16 – 8 (>, **<**) 9

Circle =, ≠, >, <.

4 + 2 (**=**, ≠) 9 – 3

2 + 3 (=, **≠**) 8 – 4

12 – 6 (>, **<**) 2 + 7

6 – 2 (>, **<**) 1 + 4

page 20 (twenty)

SELF TEST 3:

CONCEPT(S): number order, operation symbols, graphs, measurements, money

TEACHER GOAL(S): To teach the children To learn to check their progress periodically.

MATERIALS/MANIPULATIVES:
pencils, ruler

TEACHING PAGE 21:

Turn to page 21. Read the directions with the children. Be sure they understand what they are to do. You may repeat the directions but give no other help. Do not have the children check their own work. Check it as soon as you can and go over it with each child. Show him where he did well and where he needs extra help.

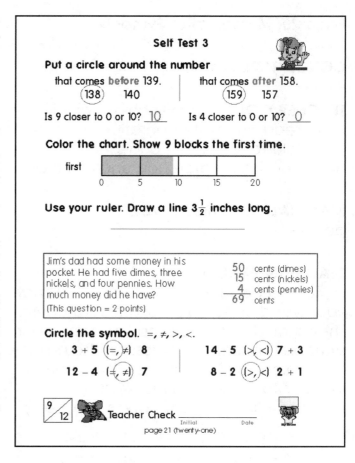

IV. PART FOUR

Page 22: Subtraction

CONCEPT(S): subtraction facts

TEACHER GOAL(S): To teach the children
To review subtraction facts for 10's
through 18's.

MATERIALS/MANIPULATIVES:
pencils, subtraction fact cards, objects
for counting

TEACHING PAGE 22:

Turn to page 22. Read the directions
with the students. Students may use objects for counting but not their fact cards
to complete this page. They may use the
answer key to check their own work.
Those students who have not mastered
their facts should spend some time
reviewing their fact cards at the end of
this lesson.

IV. Part Four

Write the answer.

12	11	15	17	13	12	11
−5	−6	−6	−8	−8	−4	−3
7	5	9	9	5	8	8

13	11	15	10	15	13	12
−6	−7	−7	−3	−9	−4	−6
7	4	8	7	6	9	6

11	18	10	14	10	15	13
−9	−9	−7	−5	−4	−8	−7
2	9	3	9	6	7	6

10	13	12	14	13	10	11
−5	−9	−7	−6	−5	−5	−4
5	4	5	8	8	5	7

11	12	17	12	14	10	12
−8	−8	−9	−9	−7	−6	−3
3	4	8	3	7	4	9

11	16	14	10	16	16	14
−5	−7	−8	−9	−9	−8	−9
6	9	6	1	7	8	5

page 22 (twenty-two)

Page 23: Place Value

CONCEPT(S): ones' place, tens' place

TEACHER GOAL(S): To teach the children
To recognize numbers in ones' place and tens' place, and
To write problems with numbers lined up in the correct place columns.

MATERIALS/MANIPULATIVES:
pencils

TEACHING PAGE 23:

Turn to page 23. Write several three digit numbers on the board and have the students identify which numbers are in the ones' place, the tens' place, and the hundreds' place. Tell the children that you are going to dictate eight vertical (up and down) addition problems to them. They must line the numbers up correctly, use the right symbol, and draw a line between the problem and the answer. Dictate:

15	26	5	2
+ 2	+ 3	+ 42	+ 31
17	29	47	33

23	50	42	31
15	19	20	22
+ 71	+ 30	+ 11	+ 63
109	99	73	116

Read the directions to the next three exercises with the children and have them complete the page.

Listen and write.

15 + 2 17	26 + 3 29	5 + 42 47	2 + 31 33
23 15 + 71 109	50 19 + 30 99	42 20 + 11 73	31 22 + 63 116

Circle the number in the ones' place.

23 145 67 93 105 8 116 43

Circle the number in the tens' place.

159 83 112 62 50 103 89 15

Count by 10.

10 20 30 40 50 60 70 80 90

page 23 (twenty-three)

Pages 24 and 25: Subtraction

CONCEPT(S): subtraction facts

TEACHER GOAL(S): To teach the children
To subtract one-digit numbers from two-digit numbers, and
To subtract two-digit numbers from two-digit numbers.

MATERIALS/MANIPULATIVES:
pencils, objects for counting

TEACHING PAGES 24 and 25:

Turn to pages 24 and 25. Point to the illustration at the top of page 24. Explain to the children that they should follow the same steps in subtraction that they follow in addition. They should subtract the ones' place *first* and the tens' place *second*. Work the first row of problems with the students and then have them read each problem aloud with the answer. Allow the students to complete that section. Point to the second illustration at the middle of page 24 and have the students subtract the ones' column first and then the tens' column. Work the first row of problems with the students and then have them read each problem aloud with the answer. Allow the students to complete that section. Point to the illustration at the middle of page 25 and have the students subtract the ones' column and then the tens' column. Work the first row of problems with the students and them have them read each problem aloud with the answer. Allow the students to complete that section. When all exercises on the two pages are complete, select certain problems and have the students illustrate these problems with objects for counting. *It is important that the students understand not just the procedure, but the concept of how many numbers they started with, how many they took away, and how many are left.*

Write the answer.

Subtract tens from tens. $\begin{array}{r} 27 \\ -\ 2 \\ \hline 25 \end{array}$ Subtract ones from ones.

$2 - (\text{nothing}) = 2 \longrightarrow \qquad \longleftarrow 7 - 2 = 5$

$\begin{array}{r} 27 \\ -\ 2 \\ \hline 25 \end{array}$	$\begin{array}{r} 38 \\ -\ 6 \\ \hline 32 \end{array}$	$\begin{array}{r} 49 \\ -\ 9 \\ \hline 40 \end{array}$	$\begin{array}{r} 65 \\ -\ 3 \\ \hline 62 \end{array}$	$\begin{array}{r} 94 \\ -\ 2 \\ \hline 92 \end{array}$	$\begin{array}{r} 18 \\ -\ 5 \\ \hline 13 \end{array}$
$\begin{array}{r} 57 \\ -\ 3 \\ \hline 54 \end{array}$	$\begin{array}{r} 79 \\ -\ 5 \\ \hline 74 \end{array}$	$\begin{array}{r} 88 \\ -\ 7 \\ \hline 81 \end{array}$	$\begin{array}{r} 68 \\ -\ 4 \\ \hline 64 \end{array}$	$\begin{array}{r} 49 \\ -\ 8 \\ \hline 41 \end{array}$	$\begin{array}{r} 33 \\ -\ 1 \\ \hline 32 \end{array}$
$\begin{array}{r} 45 \\ -\ 3 \\ \hline 42 \end{array}$	$\begin{array}{r} 63 \\ -\ 3 \\ \hline 60 \end{array}$	$\begin{array}{r} 78 \\ -\ 6 \\ \hline 72 \end{array}$	$\begin{array}{r} 94 \\ -\ 3 \\ \hline 91 \end{array}$	$\begin{array}{r} 35 \\ -\ 3 \\ \hline 32 \end{array}$	$\begin{array}{r} 24 \\ -\ 2 \\ \hline 22 \end{array}$

Subtract tens from tens. $\begin{array}{r} 30 \\ -10 \\ \hline 20 \end{array}$ Subtract ones from ones.

$3 - 1 = 2 \longrightarrow \qquad \longleftarrow 0 - 0 = 0$

$\begin{array}{r} 30 \\ -10 \\ \hline 20 \end{array}$	$\begin{array}{r} 40 \\ -20 \\ \hline 20 \end{array}$	$\begin{array}{r} 70 \\ -30 \\ \hline 40 \end{array}$	$\begin{array}{r} 90 \\ -40 \\ \hline 50 \end{array}$	$\begin{array}{r} 60 \\ -20 \\ \hline 40 \end{array}$	$\begin{array}{r} 50 \\ -10 \\ \hline 40 \end{array}$
$\begin{array}{r} 90 \\ -30 \\ \hline 60 \end{array}$	$\begin{array}{r} 20 \\ -10 \\ \hline 10 \end{array}$	$\begin{array}{r} 40 \\ -30 \\ \hline 10 \end{array}$	$\begin{array}{r} 80 \\ -40 \\ \hline 40 \end{array}$	$\begin{array}{r} 70 \\ -50 \\ \hline 20 \end{array}$	$\begin{array}{r} 40 \\ -10 \\ \hline 30 \end{array}$

page 24 (twenty-four)

$\begin{array}{r} 90 \\ -50 \\ \hline 40 \end{array}$	$\begin{array}{r} 70 \\ -40 \\ \hline 30 \end{array}$	$\begin{array}{r} 50 \\ -20 \\ \hline 30 \end{array}$	$\begin{array}{r} 90 \\ -10 \\ \hline 80 \end{array}$	$\begin{array}{r} 60 \\ -30 \\ \hline 30 \end{array}$	$\begin{array}{r} 40 \\ -40 \\ \hline 00 \end{array}$
$\begin{array}{r} 80 \\ -30 \\ \hline 50 \end{array}$	$\begin{array}{r} 60 \\ -10 \\ \hline 50 \end{array}$	$\begin{array}{r} 90 \\ -70 \\ \hline 20 \end{array}$	$\begin{array}{r} 70 \\ -20 \\ \hline 50 \end{array}$	$\begin{array}{r} 80 \\ -10 \\ \hline 70 \end{array}$	$\begin{array}{r} 90 \\ -60 \\ \hline 30 \end{array}$

Subtract tens from tens. $\begin{array}{r} 53 \\ -31 \\ \hline 22 \end{array}$ Subtract ones from ones.

$5 - 3 = 2 \longrightarrow \qquad \longleftarrow 3 - 1 = 2$

$\begin{array}{r} 53 \\ -31 \\ \hline 22 \end{array}$	$\begin{array}{r} 87 \\ -41 \\ \hline 46 \end{array}$	$\begin{array}{r} 88 \\ -16 \\ \hline 72 \end{array}$	$\begin{array}{r} 85 \\ -51 \\ \hline 34 \end{array}$	$\begin{array}{r} 79 \\ -18 \\ \hline 61 \end{array}$	$\begin{array}{r} 99 \\ -17 \\ \hline 82 \end{array}$
$\begin{array}{r} 92 \\ -20 \\ \hline 72 \end{array}$	$\begin{array}{r} 75 \\ -30 \\ \hline 45 \end{array}$	$\begin{array}{r} 77 \\ -24 \\ \hline 53 \end{array}$	$\begin{array}{r} 56 \\ -24 \\ \hline 32 \end{array}$	$\begin{array}{r} 88 \\ -17 \\ \hline 71 \end{array}$	$\begin{array}{r} 95 \\ -35 \\ \hline 60 \end{array}$
$\begin{array}{r} 93 \\ -10 \\ \hline 83 \end{array}$	$\begin{array}{r} 98 \\ -25 \\ \hline 73 \end{array}$	$\begin{array}{r} 75 \\ -25 \\ \hline 50 \end{array}$	$\begin{array}{r} 88 \\ -13 \\ \hline 75 \end{array}$	$\begin{array}{r} 83 \\ -42 \\ \hline 41 \end{array}$	$\begin{array}{r} 77 \\ -15 \\ \hline 62 \end{array}$
$\begin{array}{r} 59 \\ -42 \\ \hline 17 \end{array}$	$\begin{array}{r} 73 \\ -51 \\ \hline 22 \end{array}$	$\begin{array}{r} 65 \\ -53 \\ \hline 12 \end{array}$	$\begin{array}{r} 97 \\ -14 \\ \hline 83 \end{array}$	$\begin{array}{r} 88 \\ -22 \\ \hline 66 \end{array}$	$\begin{array}{r} 86 \\ -72 \\ \hline 14 \end{array}$

page 25 (twenty-five)

Mathematics 109 Teacher Notes

Page 26: Addition

CONCEPT(S): checking addition problems

TEACHER GOAL(S): To teach the children To check addition problems for the correct answer.

MATERIALS/MANIPULATIVES: pencils

TEACHING PAGE 26:

Turn to page 26 and read the directions at the top of the page. Ask the children if they remember how to check their addition answers to be sure that they are right. Remind them that they always add the ones' column first whether they are adding down or adding up. Tell the children to complete the page by adding down and then to check their answers by adding up. Monitor the students work to be sure they are adding up and not just copying answers from adding down. Tell them that when their answers are not the same, they know they have made a mistake and they should go back and correct their error.

Add down. Add up.

15	6	7	8	15	15
5	6	3	2	6	8
5	0	3	4	6	2
+ 5	+ 0	+ 1	+ 2	+ 3	+ 5
15	6	7	8	15	15

37	66	99	92	56	69
34	61	96	72	32	29
+ 3	+ 5	+ 3	+ 20	+ 24	+ 40
37	66	99	92	56	69

88	90	88	78	74	88
56	80	66	17	73	72
+ 32	+ 10	+ 22	+ 61	+ 1	+ 16
88	90	88	78	74	88

89	97	99	128	157	109
16	34	72	35	51	26
20	52	06	22	40	51
+ 53	+ 11	+ 21	+ 71	+ 66	+ 32
89	97	99	128	157	109

page 26 (twenty-six)

Page 27: Subtraction

CONCEPT(S): subtraction of one and two digit numbers

TEACHER GOAL(S): To teach the children To review subtraction of one- and two-digit numbers.

MATERIALS/MANIPULATIVES:
pencils, crayons

TEACHING PAGE 27:

Turn to page 27 and read the directions at the top of the page. Ask the children to describe the difference between the first row of problems and the other problems on the page. (First row is subtraction facts.) Have the children look at the other problems and describe how they will solve them (ones' column first, tens' column second). Tell the students that when they have completed the page they may color the picture at the bottom of the page.

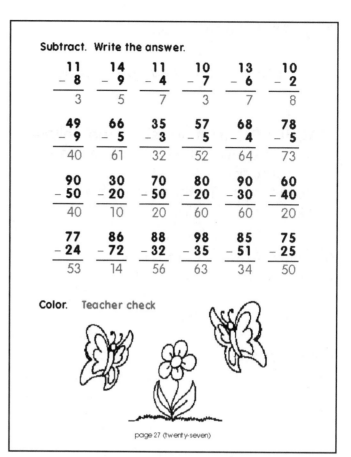

Subtract. Write the answer.

11	14	11	10	13	10
− 8	− 9	− 4	− 7	− 6	− 2
3	5	7	3	7	8

49	66	35	57	68	78
− 9	− 5	− 3	− 5	− 4	− 5
40	61	32	52	64	73

90	30	70	80	90	60
− 50	− 20	− 50	− 20	− 30	− 40
40	10	20	60	60	20

77	86	88	98	85	75
− 24	− 72	− 32	− 35	− 51	− 25
53	14	56	63	34	50

Color.   Teacher check

page 27 (twenty-seven)

Page 28: Number Value

CONCEPT(S): number value and number sense

TEACHER GOAL(S): To teach the children
To understand the relative value of a number, and
To decide whether the answer to a problem is sensible.

MATERIALS/MANIPULATIVES:
pencils, paper, orange crayon

TEACHING PAGE 28:
Ask the children to write fifteen numbers between *0* and *99* in a column on a piece of paper. Tell them that there are fifteen questions in today's lesson and that they will use these numbers to answer the questions. Talk to the children about the word *nonsense*. Talk about the fact that *nonsense* means that something makes *no sense* or is *not sensible.* Tell them that it is important when they do a problem that their answer is always a sensible answer.

Turn to page 28. Have the students look at the picture and read the rhyme with them.

Ask the children whether they think Matt the cat's answer is a sensible one. Is the reason that Herb the bird cannot read because he has sunglasses on? Next go down the list of questions with the students. Have the students write their first number on the blank in the first question and then read the sentence. Their answer might be: *My friend Matthew is 69 years old.* Continue to the next question with the next number on the student's list and so on to complete the page. When the questions are all answered, have the students circle the nonsense answers with an orange crayon.

Nonsense

Herb the bird is sitting,
High up in a tree.
He thinks that he is reading,
Just like you and me.

Matt the cat says, "Nonsense!
You know that cannot be.
The shades that Herb is wearing,
Make it much too dark to see."

My friend Matthew is _____ years old.   **Teacher check**

He has _____ fingers and _____ toes.

His birthday is June _____.

He has _____ brothers _____ sisters.

Yesterday, he ate _____ apples for lunch.

Sue has _____ pet dogs.

Each dog has _____ tail(s).

There are _____ days in the week.

The mailperson delivered _____ packages to Mary's house yesterday.

Betty is having a birthday party. She has invited _____ people.

Mark has _____ pennies in his pockets.

Laura has a dozen eggs. She fell and broke _____ of them.

Patty walked _____ blocks to friend's house.

Circle the 'nonsense' answers with orange crayon.
page 28 (twenty-eight)

SELF TEST 4:

CONCEPT(S): place value, subtraction of two-digit numbers, sensible answers

TEACHER GOAL(S): To teach the children To learn to check their progress periodically.

MATERIALS/MANIPULATIVES: pencils

TEACHING PAGE 29:

Turn to page 29. Read the directions with the children. Be sure they understand what they are to do. You may repeat the directions but give no other help. Do not have the children check their own work. Check it as soon as you can and go over it with each child. Show him where he did well and where he needs extra help.

Listen and write.
Dictate:

42	6	40	15
+ 5	+ 21	36	21
47	27	+ 12	+ 33
		88	69

Self Test 4

Listen and write. (2 pts. each exercise)

42 + 5 47	6 + 21 27	40 36 + 12 88	15 21 + 33 69

Circle the number in the

ones' place. (13) (142) (80)   tens' place. (55) (73) (104)

Write the answer.

10 - 2 8	11 - 5 6	12 - 7 5	13 - 9 4	14 - 6 8	15 - 8 7
27 - 2 25	98 - 6 92	40 - 20 20	70 - 30 40	56 - 42 14	88 - 25 63

Circle the correct answer.

Jerry has 25 fingers and 42 toes. (sensible, (nonsense))

22/27 Teacher Check _____
page 29 (twenty-nine)

V. PART FIVE

Page 30: Shapes and Fractions

CONCEPT(S): shapes and fractions

TEACHER GOAL(S): To teach the children
To recognize flat shapes, and
To divide shapes into parts and
express the parts as fractions.

MATERIALS/MANIPULATIVES:

This page requires the teacher to prepare shapes identical to the ones shown on teaching page 30. The shapes should be made from four different colors of construction paper and of the same size shown on page 30: 1) a square two inches by two inches cut into two equal parts 2) a circle with a diameter of two inches cut into four equal parts 3) a triangle with all three sides equal to two inches cut into three equal parts 4) a rectangle one inch by two inches cut into five equal parts. A set is needed for every student; also, a paper bag, pencils, and glue.

TEACHING PAGE 30:

Put the prepared pieces in a paper bag. Let the students mix them and then shake them from the bag onto a table or desk. Tell the students that they are to take the pieces and to put them together to make four different shapes. Tell them that the only hint they have are the colors. When the shapes are together, put different parts of the shapes aside and tell the students to describe the part(s) as a fraction. Ask them to describe what is left as a fraction. Do this long enough for the students to respond correctly to the question.

Turn to page 30. Begin with the square. Have the students glue one piece of the

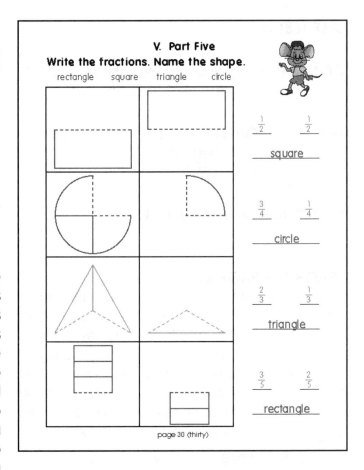

V. Part Five
Write the fractions. Name the shape.
rectangle   square   triangle   circle

page 30 (thirty)

square in one box and one piece of the square in the second box as illustrated. Tell the children to write the fractions on the lines that describe the pieces of the square ($\frac{1}{2}$, $\frac{1}{2}$). Remind the students that they should write the denominator first (the number of parts the shape is divided into) and the numerator (the number of parts being described) second. Have the children write the name of the shape on the long line below the fractions. Complete the page gluing the parts, writing the fractions, and naming the shapes.

Page 31: Fractions

CONCEPT(S): fractions and sets

TEACHER GOAL(S): To teach the children
   To divide sets into parts, and
   To express the parts as a fraction.

MATERIALS/MANIPULATIVES:
pencils, paper, objects to make up four sets, (a set of 6, a set of 9, a set of 12, a set of 15)

TEACHING PAGE 31:
   Place each one of the sets in front of the students and ask them to describe the set.

   The students will initially respond 6 of one thing, 9 of another. Explain to them that there is another way to describe each set. It may be described as one set of 6 paper clips, one set of 9 pencils, and so on. Emphasize the expression *one set of*. Have the children take the set of 6 objects and divide it into two groups of 2 and 4. Ask them how many objects are in each group. Help them to write a fraction on paper that will describe what they have done. Ask how many parts the set is divided into (6). Tell them this is the bottom number of the fraction (denominator). Ask how many objects in each group (2, 4). Tell them this is the top number (numerator) of the fraction. Ask the children to say the fractions aloud (two-sixths, four-sixths).

   Turn to page 31. Tell the children to draw six tally marks in the first box to represent the whole set. Ask them to draw two tally marks in the second box and four in the third box to represent the two groups. Have them write the fraction for each group ($\frac{2}{6}$ and $\frac{4}{6}$) on the two lines and then write the name of the objects they have used on the third line. Follow

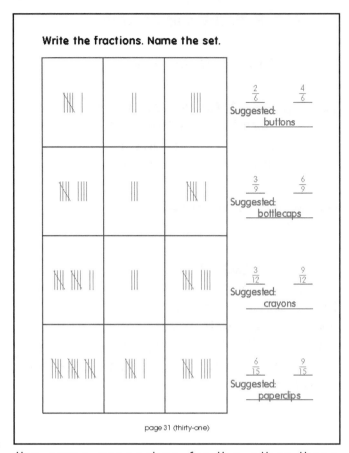

Write the fractions. Name the set.

			$\frac{2}{6}$ $\frac{4}{6}$ Suggested: buttons

page 31 (thirty-one)

the same procedure for the other three sets. Divide the set of 9 into two groups, ($\frac{3}{9}$ and $\frac{6}{9}$), the set of 12 into two groups, ($\frac{3}{12}$ and $\frac{9}{12}$), and the set of 15 into two groups, ($\frac{6}{15}$ and $\frac{9}{15}$). Be sure the children say the fractions aloud as they write them. *Note: The concept of the ratio $\frac{2}{6}$ being equal to $\frac{1}{3}$ or $\frac{3}{12}$ being equal to $\frac{1}{4}$ is not one that most early learners will grasp.*

Do not expect too much of your students. The important points for your students to understand are that a fraction means part of a whole. The bottom number tells "how many" the whole is divided into. The top number tells "how many" we are talking about.

Page 32: Patterns and Sequences

CONCEPT(S): patterns and sequences

TEACHER GOAL(S): To teach the children
To recognize a pattern, and
To tell what comes next in the
sequence.

MATERIALS/MANIPULATIVES:
pencils

TEACHING PAGE 32:
Turn to page 32 and read the directions aloud with the students. Work the first exercise with them to be sure they understand what is expected of them. Allow them to complete the page independently.

Find the pattern. Draw or write what comes next.

147 , 148 , 149	150
March , April , May	June
13 13 13 − 4 − 5 − 6	13 − 7
10 20 30	40

page 32 (thirty-two)

Page 33: Place Value, Add and Subtract

CONCEPT(S): place value for ones, tens, hundreds, addition and subtraction problems

TEACHER GOAL(S): To teach the children
To identify number place for ones, tens, and hundreds, and
To review addition and subtraction problems.

MATERIALS/MANIPULATIVES:
pencils

TEACHING PAGE 33:
Turn to page 33 and read the directions aloud to both exercises with the students. Work the first problem in the first exercise with the children to be sure they understand what is expected of them. Allow them to complete the page independently.

Write the number.

23 has a __2__ in the tens' place.

64 has a __4__ in the ones' place.

6 has a __6__ in the ones' place.

134 has a __1__ in the hundreds' place.

156 has a __5__ in the tens' place.

103 has a __3__ in the ones' place.

Add and subtract.

62	24	37	71	6	25
+ 3	+ 5	+ 52	+ 20	3	12
				+ 4	+ 31
65	29	89	91	13	68

34	48	70	50	63	89
− 2	− 5	− 30	− 40	− 21	− 43
32	43	40	10	42	46

page 33 (thirty-three)

SELF TEST 5:

CONCEPT(S): fractions, place value, addition, subtraction

TEACHER GOAL(S): To teach the children To learn to check their progress periodically.

MATERIALS/MANIPULATIVES:
pencils

TEACHING PAGE 34:
Turn to page 34. Read the directions with the children. Be sure they understand what they are to do. You may repeat the directions but give no other help. Do not have the children check their own work. Check it as soon as you can and go over it with each child. Show him where he did well and where he needs extra help.

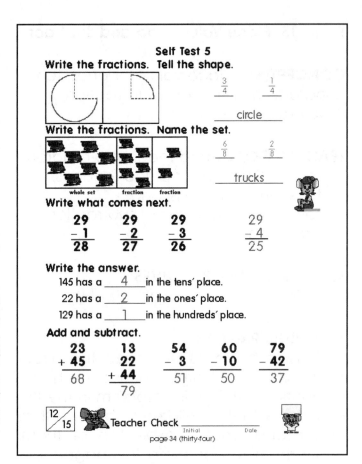

LIFEPAC TEST AND ALTERNATE TEST 109

CONCEPT(S): addition, subtraction, place value, time, number words, fractions, listening skills

TEACHER GOALS: To teach the children
To learn to check their own progress periodically.

MATERIALS/MANIPULATIVES:
pencils, rulers

TEACHING the LIFEPAC TEST:
Administer the test in at least two sessions.

Read all of the directions on each page as the children prepare to do it. Be sure that they understand what they are being asked to do. Give no help except with directions. Go over each page with the child as soon as possible after you check it so that he can see where he did well and where he needs more work.

Evaluate the tests and review areas where the children have done poorly. Review the pages and activities that stress the concepts tested.

If necessary, when the children have reviewed sufficiently, administer the Alternate LIFEPAC test. Follow the same procedures as used for the LIFEPAC Test.

LIFEPAC Test
Listen and write. (page 3)
Numbers must be written in the correct columns by place value. Problems should include plus (+) signs and lines.

Dictate:

```
  41        3       14        3
+  3      + 25      21       82
  44       28     + 50     + 11
                    85       96
```

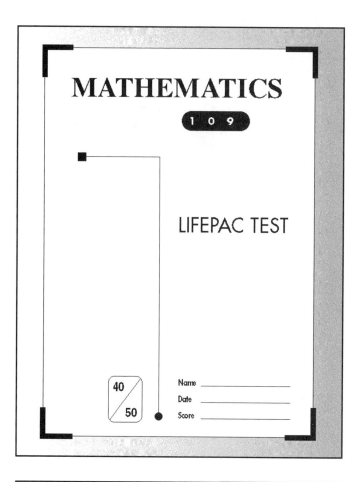

MATHEMATICS 109: LIFEPAC TEST

Complete the fact families for addition and subtraction. (2 points each)

5	3	**7**	6	**14**	14	**12**	12
+ 3	+ 5	**+ 6**	+ 7	**− 9**	− 5	**− 3**	− 9
8	8	13	13	5	9	9	3

Add or subtract.

```
  5          23
  4    64    42    12    76    87
+ 6   + 5   + 51   − 8   − 5   − 24
 15    69   116     4    71    63
```

Write the number word.

13 _____thirteen_____   33 _____thirty-three_____

Write the days in order. (3 points)

Thursday	Sunday	Tuesday	Monday	Saturday	Wednesday	Friday
5	1	3	2	7	4	6

Write the time.

_2:50_   _11:05_   _6:20_   _8:35_

page 1 (one)

Alternate LIFEPAC Test

Dictate:

```
   23        5        31        4
 + 5       + 42       18       63
  28        47       + 40     + 10
                       89       77
```

Write the answer.

Is 4 closer to 0 or 10? _0_   Is 6 closer to 0 or 10? _10_

Tammie has four dimes, three nickels, and two pennies in her hand. How much money does she have? (2 points)

40	cents (dimes)
15	cents (nickels)
_2_	cents (pennies)
57	cents

Circle =, ≠, >, <.

2 + 6 (=, ≠) 7   16 − 8 (>, <) 4 + 3

18 − 9 (=, ≠) 2 + 7   6 − 2 (>, <) 5 + 1

Listen and write. (2 points each)

41 + 3 44	3 + 25 28	14 21 + 50 85	3 82 + 11 96

Circle the correct answer.

James has a dozen eggs. He fell and broke 14 of them.

(sensible, nonsense)

page 2 (two)

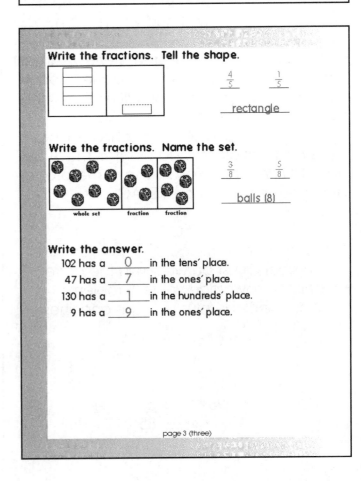

Write the fractions. Tell the shape.

$\frac{4}{5}$ $\frac{1}{5}$

rectangle

Write the fractions. Name the set.

$\frac{3}{8}$ $\frac{5}{8}$

balls (8)

whole set   fraction   fraction

Write the answer.

102 has a _0_ in the tens' place.

47 has a _7_ in the ones' place.

130 has a _1_ in the hundreds' place.

9 has a _9_ in the ones' place.

page 3 (three)

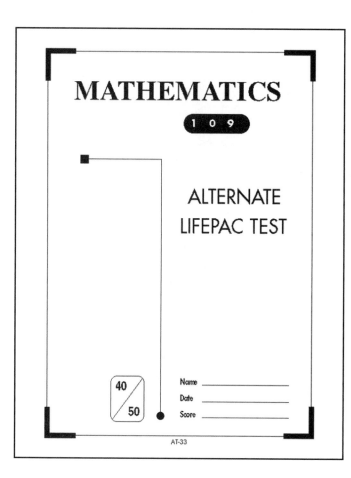

MATHEMATICS

1 0 9

ALTERNATE LIFEPAC TEST

40 / 50

Name _____
Date _____
Score _____

AT-33

MATHEMATICS 109: LIFEPAC TEST

Complete the fact families for addition and subtraction. (2 points each)

4	8	6	3	15	15	8	8
+ 8	+ 4	+ 3	+ 6	− 7	− 8	− 1	− 7
12	12	9	9	8	7	7	1

Add or subtract.

3	45				
2	30				
+ 5	73	+ 21	13	67	83
	+ 6		− 5	− 4	− 71
10	79	96	8	63	12

Write the number word.

12 _____twelve_____   22 _____twenty-two_____

Write the days in order. (3 points)

Saturday	Monday	Tuesday	Friday	Thursday	Wednesday	Sunday
7	2	3	6	5	4	1

Write the time.

4:55   9:05   8:20   1:40

AT-34

Write the answer.

Is 8 closer to 0 or 10? __10__   Is 3 closer to 0 or 10? __0__

Tammie has five dimes, two nickels, and six pennies in her hand. How much money does she have? (2 points)

50 cents (dimes)
10 cents (nickels)
6 cents (pennies)
66 cents

Circle =, ≠, >, <.

9 + 3 (**=**, ≠) 11   8 − 5 (>, **<**) 4 + 0

14 − 7 (**=**, ≠) 3 + 4   12 − 7 (>, **<**) 16 − 8

Listen and write. (2 points each)

23	5	31	4
+ 5	+ 42	18	63
28	47	+ 40	+ 10
		89	77

Circle the correct answer.

Jenny spent the day at her friends house. She left home at 9:00 PM and came home at 3:00 AM.

(sensible, **nonsense**)

AT-35

Write the fractions. Tell the shape.

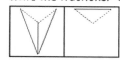

$\frac{2}{3}$ $\frac{1}{3}$

_____triangle_____

Write the fractions. Name the set.

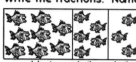

whole set   fraction   fraction

$\frac{5}{8}$ $\frac{3}{8}$

_____fish (8)_____

Write the answer.

39 has a __3__ in the tens' place.
106 has a __6__ in the ones' place.
170 has a __1__ in the hundreds' place.
5 has a __5__ in the ones' place.

AT-36

Page 1: Fun With Numbers

CONCEPT(S): purpose of LIFEPAC, objectives

TEACHER GOAL(S): To teach the children
To know what is expected of the student in the LIFEPAC, and
To write first and last names correctly in manuscript.

MATERIALS/MANIPULATIVES:
pencils

TEACHING PAGE 1:

Turn to page 1. Point to the title and the memory verse and read them aloud. Allow time for the children to look through the LIFEPAC. Write the word *OBJECTIVES* on the board and have the children find the word on the page. Explain that the objectives tell the things the students will be expected to do in the LIFEPAC. Read each one and have the children repeat as they run their fingers under the sentence from left to right. Talk about the objectives so that the children will understand what they will be doing. Have each child write his name on the line.

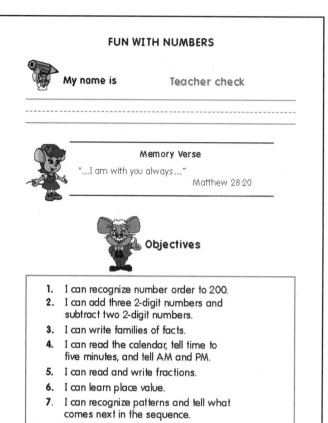

FUN WITH NUMBERS

My name is   Teacher check

Memory Verse
"...I am with you always..."
Matthew 28:20

Objectives

1. I can recognize number order to 200.
2. I can add three 2-digit numbers and subtract two 2-digit numbers.
3. I can write families of facts.
4. I can read the calendar, tell time to five minutes, and tell AM and PM.
5. I can read and write fractions.
6. I can learn place value.
7. I can recognize patterns and tell what comes next in the sequence.

page 1 (one)

I. PART ONE

Pages 2 and 3: Numbers to 200

CONCEPT(S): number order to 200, even and odd numbers, skip counting

TEACHER GOAL(S): To teach the children
   To write numbers to 200,
   To recognize even and odd numbers, and
   To skip count by 2's, 5's, and 10's.

MATERIALS/MANIPULATIVES:
pencils; yellow, green, purple, and orange crayons

TEACHING PAGES 2 and 3:
   Turn to page 2 and read the directions at the top of the page. When the children have completed this exercise, review even and odd numbers with them. Remind them that even numbers always end in *0, 2, 4, 6, 8* and that odd numbers always end in *1, 3, 5, 7, 9*. Have the children complete the page.

   Continue to page 3, and read the directions at the top of the page. When the students have completed writing the numbers, read the directions at the bottom of the page. When the children understand what they are to do, they may go on to complete the page.

I. Part One

Write numbers to 100.
Write the number in each box.

0	1	2	3	4	5	6	7	8	9
10	11	12	13	14	15	16	17	18	19
20	21	22	23	24	25	26	27	28	29
30	31	32	33	34	35	36	37	38	39
40	41	42	43	44	45	46	47	48	49
50	51	52	53	54	55	56	57	58	59
60	61	62	63	64	65	66	67	68	69
70	71	72	73	74	75	76	77	78	79
80	81	82	83	84	85	86	87	88	89
90	91	92	93	94	95	96	97	98	99
100									

0 1 2 3 4 5 6 7 8 9

Even numbers end in 0   2   4   6   8 .

Odd numbers end in 1   3   5   7   9 .

Color the even numbers from 12 to 22 purple. **Teacher check**

Color the odd numbers from 41 to 53 orange. **Teacher check**

page 2 (two)

Write numbers to 200.
Write the number in each box.

100	101	102	103	104	105	106	107	108	109
110	111	112	113	114	115	116	117	118	119
120	121	122	123	124	125	126	127	128	129
130	131	132	133	134	135	136	137	138	139
140	141	142	143	144	145	146	147	148	149
150	151	152	153	154	155	156	157	158	159
160	161	162	163	164	165	166	167	168	169
170	171	172	173	174	175	176	177	178	179
180	181	182	183	184	185	186	187	188	189
190	191	192	193	194	195	196	197	198	199
200									

Begin at 0 and count by 10's to 200.   **Teacher check**
Write a yellow X on each number.

Begin at 85 and count by 5's to 125.   **Teacher check**
Put a green circle around each number.

page 3 (three)

Page 4: Number Words, Telling Time

CONCEPT(S): facts as number words, time to 5 minutes

TEACHER GOAL(S): To teach the children
To recognize addition and subtraction facts written in words, and
To tell time to five minutes.

MATERIALS/MANIPULATIVES:
pencils

TEACHING PAGE 4:
Turn to page 4. Point to several lines in the first box and ask the students to read the words aloud. Ask the students what these are called (number sentences). Point out the capital letters and the periods. Remind the children that their answers must be written as number words. Have the students complete the first problem to be sure they understand the instructions. Go on to the second exercise. Read the directions with the students. The children should be able to complete the number word facts and the clock exercises without further help.

Add or subtract.

Six plus eight equals __fourteen__ .

Eight minus three equals __five__ .

Nine minus four equals __five__ .

Nine plus two equals __eleven__ .

Three plus two equals __five__ .

Sixteen minus eight equals __eight__ .

Seventeen minus eight equals __nine__ .

Seven plus seven equals __fourteen__ .

Four plus nine equals __thirteen__ .

Six minus zero equals __six__ .

Eleven minus five equals __six__ .

Three plus four equals __seven__ .

Write the time.

8:30   7:15   8:45   4:55

page 4 (four)

Page 5: Number Lines and Number Symbols

CONCEPT(S): numbers to the nearest 10, number symbols (>, <, =, ≠)

TEACHER GOAL(S): To teach the children
To show the multiple of ten that a number is closest to, and
To use number symbols (>, <, =, ≠) correctly.

MATERIALS/MANIPULATIVES:
pencils

TEACHING PAGE 5:
Turn to page 5. Ask the students to count by *10* as they point to the number line.

Talk about the small marks on the line and ask the children what they stand for. Have the children count from *1* to *30* on the line as they point to each small mark.

Direct the children's attention to the first question. Ask them to locate the number *8* on the number line and then to answer the question. Allow the children to complete the first exercise in this manner. Write the symbols >, <, =, ≠ on the board and discuss their meaning. Remind the students that the open part of the > sign is always toward the larger number. Read the directions to the second exercise on the page. Direct the children's attention to the one hundred words. Have them say the words aloud and write them in symbols before completing the exercise. Have them complete the page.

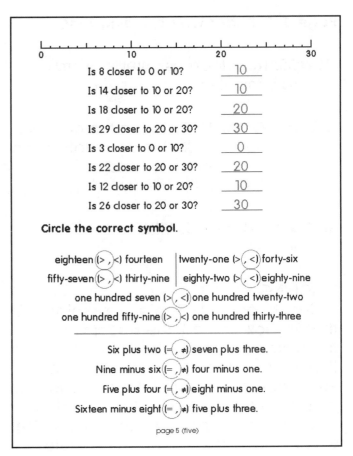

Is 8 closer to 0 or 10?   10
Is 14 closer to 10 or 20?   10
Is 18 closer to 10 or 20?   20
Is 29 closer to 20 or 30?   30
Is 3 closer to 0 or 10?   0
Is 22 closer to 20 or 30?   20
Is 12 closer to 10 or 20?   10
Is 26 closer to 20 or 30?   30

Circle the correct symbol.

eighteen (>, <) fourteen | twenty-one (>, <) forty-six
fifty-seven (>, <) thirty-nine | eighty-two (>, <) eighty-nine
one hundred seven (>, <) one hundred twenty-two
one hundred fifty-nine (>, <) one hundred thirty-three

Six plus two (=, ≠) seven plus three.
Nine minus six (=, ≠) four minus one.
Five plus four (=, ≠) eight minus one.
Sixteen minus eight (=, ≠) five plus three.

page 5 (five)

Page 6: Estimation and Skip Counting

CONCEPT(S): estimation, skip counting

TEACHER GOAL(S): To teach the children
To estimate groups of objects, and
To learn to count quickly by grouping numbers by 2's, 5's, and 10's.

MATERIALS/MANIPULATIVES:
pencils, objects for counting, (pennies, beans, buttons, bottle caps)

TEACHING PAGE 6:
Place a group of objects (about twenty) in front of the students. Have them count the set by 1's, taking one object at a time (1-2-3 and so on). Ask them if they remember an easier way to count the set. Talk to them about skip counting. Have the children count the set again by 2's, taking two objects at a time (2-4-6 and so on). Ask which method they think is easier and quicker.

Turn to page 6. Read the title, the instructions, and the words at the top of each column with the students. Complete the page in the following manner. Select a set of objects and have the children write the name in the first column (buttons). Ask them how they would like to skip count this set, (by 2's, by 5's, by 10's), and write that in the next column. Have the students estimate the number of objects in the set, and write the number under *estimate*. Next, have them count (by 2's, 5's, or 10's) and write the count under *how many*. (To count 32 by fives, count 6 sets of fives plus 2 ones.) Finally, have them circle the correct operation symbol (>,<). Use a variety of objects and numbers of objects to make the exercise more interesting. Have the students complete the addition wheel at the bottom of the page. (6 + 5 and so on around the wheel.)

Suggested answer — **Easy Counting** — Teacher check

Name	By 2's,5's,10's	Estimate	Circle	How many
buttons	5's	43	(>,<)	32
			(>,<)	
			(>,<)	
			(>,<)	
			(>,<)	
			(>,<)	
			(>,<)	
			(>,<)	

Addition wheel (6 +): 8, 11, 2, 5, 12, 6, 9, 15, 6 +, 14, 8, 4, 10, 3, 7, 9, 13

page 6 (six)

Mathematics 110 Teacher Notes

SELF TEST 1:

CONCEPT(S): numbers to 200, number words and symbols, numbers to the nearest 10

TEACHER GOAL(S): To teach the children To learn to check their progress periodically.

MATERIALS/MANIPULATIVES: pencils

TEACHING PAGE 7:

Turn to page 7. Read the directions with the children. Be sure they understand what they are to do. You may repeat the directions but give no other help. Do not have the children check their own work. Check it as soon as you can and go over it with each child. Show him where he did well and where he needs extra help.

Self Test 1

Write the missing numbers. (each row 1 point)

0	1	2	3	4	5	6	7	8	9
90	91	92	93	94	95	96	97	98	99
(140)	141	(142)	143	(144)	145	(146)	147	(148)	149

Even numbers end in __0_ _2_ _4_ _6_ _8_.
Circle the even numbers in the third row. (5 points)

Add or subtract.

Six plus five equals ___eleven___.
Eight minus five equals ___three___.
Nine plus four equals ___thirteen___.
Fifteen minus seven equals ___eight___.

Write the answer.

Is 8 closer to 0 or 10? _10_   Is 23 closer to 20 or 30? _20_
Is 17 closer to 10 or 20 _20_   Is 12 closer to 10 or 20? _10_

Circle the correct answer.

seventeen (>, <) fourteen
one hundred twelve (>, <) one hundred twenty
seven plus eight (=, ≠) nine plus six

19/24 Teacher Check _____
        Initial   Date
page 7 (seven)

II. PART TWO

Page 8: Addition, Counting Money

CONCEPT(S): one- and two-digit addition problems, converting cents to coins

TEACHER GOAL(S): To teach the children
To solve addition problems using one and two-digit numbers in columns, and
To convert pennies, nickels, and dimes to cents.

MATERIALS/MANIPULATIVES:
pencils, pennies, nickels, dimes

TEACHING PAGE 8:

Turn to page 8. Talk to the children about ones' place and tens' place. Ask which place is always added first (ones'). Read the directions at the bottom of the page. Tell the children that they should count the coins and then write the amount of cents that they have. Children may use actual coins to help solve these problems. Allow the children time to complete the page.

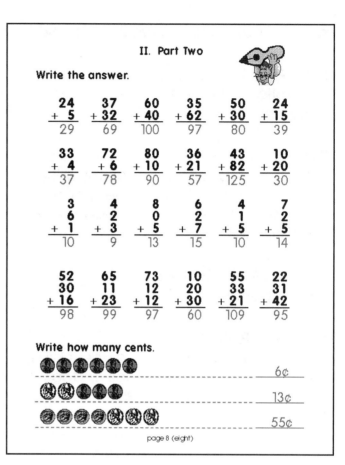

Page 9: Subtraction, Counting Money

CONCEPT(S): one- and two-digit subtraction problems, converting coins to cents

TEACHER GOAL(S): To teach the children
To solve subtraction problems using one- and two-digit numbers, and
To convert cents to pennies, nickels, and dimes.

MATERIALS/MANIPULATIVES:
pencils, pennies, nickels, dimes

TEACHING PAGE 9:
Turn to page 9. Talk to the children about ones' place and tens' place. Ask which place is always subtracted first (ones'). Read the directions at the bottom of the page. Tell the children that they should look at the cents and then write the amount of coins that they have. Children may use actual coins to help solve these problems. *Answers to the distribution of coins may vary.* Allow the children time to complete the page.

Write the answer.

73	87	70	85	70	43
− 2	−35	−50	− 4	−30	−21
71	52	20	81	40	22

63	48	30	75	59	90
− 2	− 6	−20	−13	−36	−40
61	42	10	62	23	50

60	46	92	66	78	70
−40	−24	−70	− 3	− 5	−50
20	22	22	63	73	20

60	45	56	50	59	59
−30	− 2	−42	−20	− 4	−35
30	43	14	30	55	24

65	90	78	57	80	79
− 3	−10	− 2	−26	−20	−25
62	80	76	31	60	54

Write how many dimes. nickels. pennies.

	dimes.	nickels.	pennies.
24¢ Suggested answer	2		4
48¢	4	1	3
67¢	6	1	2

page 9 (nine)

Page 10: Number Order

CONCEPT(S): number order from one hundred to two hundred

TEACHER GOAL(S): To teach the children
To arrange number words from one hundred to two hundred in number order, and
To recognize ordinal numbers as words from first to sixth.

MATERIALS/MANIPULATIVES:
pencils; green, blue, yellow, purple, red, and orange crayons

TEACHING PAGE 10:

Turn to page 10. Talk about the illustrations on the page and read the rhyme with the students. Point out the words one hundred and have the children read the number words aloud. (Ask them if the numbers are in number order. Tell them to write the number words in number symbols in the boxes.)

Point to the words *first through sixth* on the umbrella. Tell the children to arrange the number symbols they have written in order from first through sixth on the lines below the umbrella. Then have the children color the umbrella using the same colors as the corresponding number boxes.

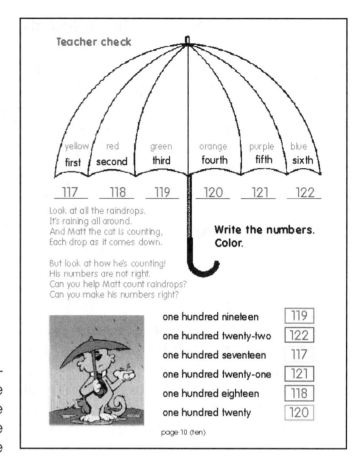

189

Page 11: Families of Facts, Measurements

CONCEPT(S): families of facts for addition and subtraction, measuring inches

TEACHER GOAL(S): To teach the children
   To write families of facts for addition and subtraction, and
   To draw lines to the half inch.

MATERIALS/MANIPULATIVES:
pencils, objects for counting, ruler measuring up to 6 inches

TEACHING PAGE 11:

Turn to page 11 and read the directions. Point to the first box and have the students say the numbers aloud. Tell them that these numbers make a family of addition and subtraction facts because they are all related. Work these problems with the students helping them fill in the missing numbers. Students who have not mastered their facts and are having difficulty relating the numbers should make sets using objects for counting. *It is easier for children to remember facts when they understand that once they have learned one fact they actually know four facts.* Have the children use their rulers to draw the measurements at the bottom of the page.

Write the family of facts.

2, 3, 5		4, 2, 6		8, 7, 15	
2 + 3 ‾5‾	3 + 2 ‾5‾	4 + 2 ‾6‾	2 + 4 ‾6‾	8 + 7 ‾15‾	7 + 8 ‾15‾
5 − 3 ‾2‾	5 − 2 ‾3‾	6 − 2 ‾4‾	6 − 4 ‾2‾	15 − 7 ‾8‾	15 − 8 ‾7‾

5, 3, 8		6, 5, 11		9, 4, 13	
5 + 3 ‾8‾	5 + 3 ‾8‾	6 + 5 ‾11‾	5 + 6 ‾11‾	9 + 4 ‾13‾	4 + 9 ‾13‾
8 − 5 ‾3‾	8 − 3 ‾5‾	11 − 5 ‾6‾	11 − 6 ‾5‾	13 − 9 ‾4‾	13 − 4 ‾9‾

Draw lines. Use a ruler.

6 inches ——————————————

$3\frac{1}{2}$ inches —————————

$1\frac{1}{2}$ inches ———

page 11 (eleven)

Page 12: Plane and Solid Shapes

CONCEPT: naming plane and solid shapes

TEACHER GOAL(S): To teach the children
To name plane and solid shapes, and
To write number words.

MATERIALS/MANIPULATIVES:
pencils, crayons

TEACHING PAGE 12:

Turn to page 12 and tell the students to complete the matching part of the exercise.

Have the children select the flat shapes and outline each one using the color of the name box. Ask the children to find the solid shapes and talk to them about the different way these shapes are illustrated. Point out the extra lines that are drawn to give the solid shapes sides, a top, and a bottom. Tell the children to use their crayons to draw over these lines (to develop their sense of dimension). Have the children color the inside of each shape using a lighter tone of the color so the outline stands out. When the coloring is complete, tell the children to write the number words at the bottom of the page.

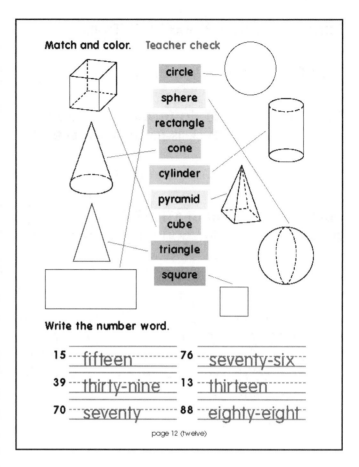

Match and color.　Teacher check

circle
sphere
rectangle
cone
cylinder
pyramid
cube
triangle
square

Write the number word.

15 fifteen　　76 seventy-six
39 thirty-nine　13 thirteen
70 seventy　　88 eighty-eight

page 12 (twelve)

Page 13: Reading and Math Skills

CONCEPT(S): solving story problems

TEACHER GOAL(S): To teach the children
  To apply math skills to story problems, and
  To read for comprehension.

MATERIALS/MANIPULATIVES:
pencils, objects for counting, crayons, clock

TEACHING PAGE 13:
Turn to page 13. Students should be encouraged to read the problems on this page and to solve them independently. They should show the work that they have done to solve the problem on the LIFEPAC page and they should write the correct answer on the line, (Example: *12 + 23 + 14 = 49* *49* rocks). *Students should develop the habit of showing work and labeling answers as early as possible.* The exercises on this page are for math skills *and* reading comprehension. Students who are having difficulty solving a problem should read the paragraph aloud, use objects for counting, and be directed to work each problem one step at a time.

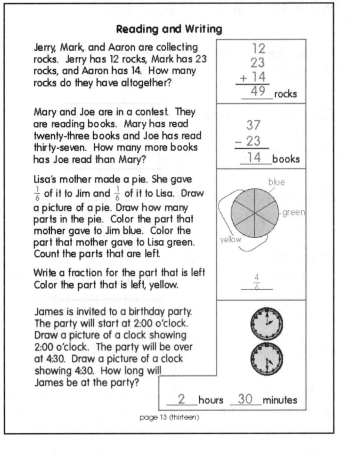

Reading and Writing

Jerry, Mark, and Aaron are collecting rocks. Jerry has 12 rocks, Mark has 23 rocks, and Aaron has 14. How many rocks do they have altogether?

$$\begin{array}{r} 12 \\ 23 \\ +14 \\ \hline 49 \end{array} \text{ rocks}$$

Mary and Joe are in a contest. They are reading books. Mary has read twenty-three books and Joe has read thirty-seven. How many more books has Joe read than Mary?

$$\begin{array}{r} 37 \\ -23 \\ \hline 14 \end{array} \text{ books}$$

Lisa's mother made a pie. She gave $\frac{1}{6}$ of it to Jim and $\frac{1}{6}$ of it to Lisa. Draw a picture of a pie. Draw how many parts in the pie. Color the part that mother gave to Jim blue. Color the part that mother gave to Lisa green. Count the parts that are left.

Write a fraction for the part that is left. Color the part that is left, yellow.

blue
green
yellow

$\frac{4}{6}$

James is invited to a birthday party. The party will start at 2:00 o'clock. Draw a picture of a clock showing 2:00 o'clock. The party will be over at 4:30. Draw a picture of a clock showing 4:30. How long will James be at the party?

_2_ hours _30_ minutes

page 13 (thirteen)

Page 14: Patterns

CONCEPT(S): patterns in facts

TEACHER GOAL(S): To teach the children
To recognize patterns in addition
and subtraction facts.

MATERIALS/MANIPULATIVES:
pencils

TEACHING PAGE 14:
Turn to page 14 and read the directions at the top of the page with the children.

Ask them to look at the facts in the first box and to write the answers to the facts. Ask the children if they see a pattern in the way the facts are written. Ask them to describe the pattern. Point out that each time one more is added to *3*, the answer is also one number larger. Ask the children what fact comes next in the pattern and tell them to write the fact in the box. Have the children continue in this manner to complete the page. Be sure that they are reading the signs correctly. Tell them to write the numbers in the sequences at the bottom of the page.

Write the answer. Write the next fact.

3 +1 — 4	3 +2 — 5	3 +3 — 6	3 +4 — 7	9 −4 — 5	9 −5 — 4	9 −6 — 3	9 −7 — 2
8 +3 — 11	8 +4 — 12	8 +5 — 13	8 +6 — 14	12 −5 — 7	12 −6 — 6	12 −7 — 5	12 −8 — 4
6 +4 — 10	6 +5 — 11	6 +6 — 12	6 +7 — 13	15 −6 — 9	15 −7 — 8	15 −8 — 7	15 −9 — 6
2 +5 — 7	2 +6 — 8	2 +7 — 9	2 +8 — 10	7 −1 — 6	7 −2 — 5	7 −3 — 4	7 −4 — 3

Write the next number word.
nine   ten   eleven   twelve   <u>thirteen</u>

Write the missing numbers.
20 <u>25 30</u> 35 <u>40 45</u> 50 <u>55 60</u> 65 <u>70</u>
130 <u>135 140</u> 145 <u>150 155</u> 160 <u>165</u> 170 <u>175 180</u>

page 14 (fourteen)

SELF TEST 2:

CONCEPT(S): addition, subtraction, cents, shapes, fractions, families of facts

TEACHER GOAL(S): To teach the children
To learn to check their progress periodically.

MATERIALS/MANIPULATIVES:
pencils, crayons

TEACHING PAGE 15:

Turn to page 15. Read the directions with the children. Be sure they understand what they are to do. You may repeat the directions but give no other help. Do not have the children check their own work. Check it as soon as you can and go over it with each child. Show him where he did well and where he needs extra help.

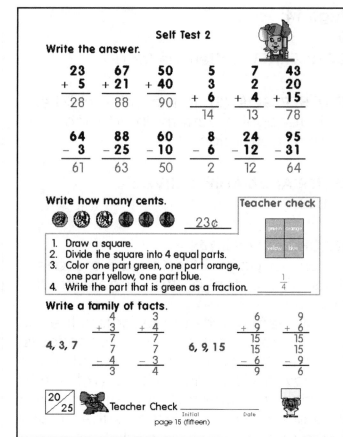

III. PART THREE

Pages 16 and 17: Following Directions

CONCEPT(S): directions - north, south, east, west; and reading to follow directions

TEACHER GOAL(S): To teach the children
To learn about the directions of north, south, east, and west; and
To learn to read a chart or map.

MATERIALS/MANIPULATIVES:
pencils

TEACHING PAGES 16 and 17:
Turn to page 16. Talk to the children about the directions of north, east, south, and west and how they are used. Ask them to look at the illustration at the top the of page. Tell the children to turn their desks so that the arrow that is marked north is pointing in the direction of north in the classroom. Tell the children to point out the directions of north, east, south, and west in the classroom. Explain to the children that the lines on page 16 represent a map. Help them find *Home* on the map. Tell them to begin at *Home* and ask where they would need to move their fingers to go north, east, south, or west from *Home*.

Turn to page 17. Ask the children to read the names of the animals written along the outside margins of the page and then to locate each one on the map. Read the instructions at the top of the page. Read the first instruction: *Start at Home*. Have the children put their fingers at *Home* on the map and then read the next instruction: *Go north two blocks*. Ask what they found (a duck). Tell the children to write *duck* on the line. Proceed in this manner (starting at Home for the first *six* instructions and as

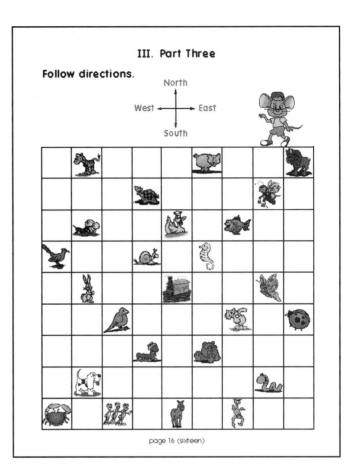

page 16 (sixteen)

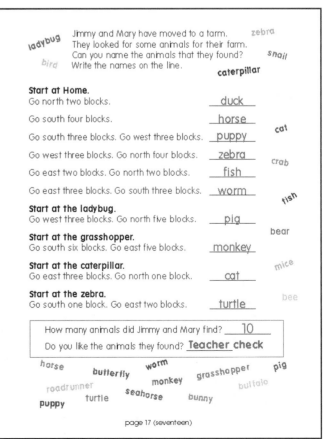

page 17 (seventeen)

instructed for the next *four*) until the children have a list of ten animals. Have them complete the page by answering the two questions at the bottom of the page.

Page 18: Graphs, Families of Facts

CONCEPT(S): data on a graph, families of facts, (addition and subtraction)

TEACHER GOAL(S): To teach the children
To post data on a graph, and
To write families of facts for addition and subtraction.

MATERIALS/MANIPULATIVES:
pencils, crayons

TEACHING PAGE 18:
Turn to page 18 and read the first set of instructions. Have the children identify and name the objects along the side of the graph (flat shapes: triangle, circle, rectangle, square). Tell them to read the numbers along the bottom of the graph. Remind the children that this is like a number line. Ask the children what they should do to show 10 triangles on the chart (color each block to the number 10). Have them complete the chart. Read the next set of directions. The students should be able to complete this exercise independently.

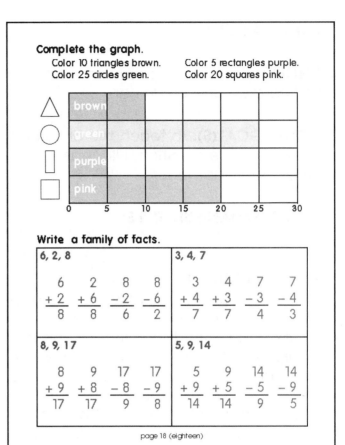

Page 19: Addition

CONCEPT(S): addition problems — one- and two-digit and columnar

TEACHER GOAL(S): To teach the children To review the addition skills that they have learned.

MATERIALS/MANIPULATIVES: pencils

TEACHING PAGE 19:

Turn to page 19. Review ones' place and tens' place. Read the directions for both exercises with the children. The last exercise will determine whether the students understand the correct way to write a vertical (up and down) addition problem. Let them complete the exercise even if it is done incorrectly. Then, review number place and have them rewrite the problem using the proper procedure.

Write the answer.

58	43	17	14	50	60
+ 11	+ 25	+ 12	+ 23	+ 20	+ 13
69	68	29	37	70	73

56	92	80	63	19	12
+ 3	+ 5	+ 14	+ 21	+ 40	+ 53
59	97	94	84	59	65

2	2	4	3	5	7
7	3	5	6	2	2
+ 1	+ 6	+ 3	+ 8	+ 9	+ 4
10	11	12	17	16	13

32	42	3	12	62	2
51	6	60	10	3	4
+ 75	+ 11	+ 52	+ 5	+ 21	+ 13
158	59	115	27	86	19

Write the problems in columns. Add.

23 + 2 + 31 =	40 + 15 + 3 =
23	40
2	15
+ 31	+ 3
56	58

page 19 (nineteen)

Page 20: Subtraction

CONCEPT(S): subtraction problems — one- and two-digit

TEACHER GOAL(S): To teach the children To review the subtraction skills that they have learned.

MATERIALS/MANIPULATIVES: pencils

TEACHING PAGE 20:

Turn to page 20. Review ones' place and tens' place. Read the directions for both exercises with the children. The last exercise will determine whether the students understand the correct way to write a vertical (up and down) subtraction problem. Let them complete the exercise even if it is done incorrectly. Then, review number place and have them rewrite the problem using the proper procedure.

Write the answer.

28	56	30	86	42	15
$-\ 3$	$-\ 4$	-20	-25	-11	-12
25	52	10	61	31	3

59	63	97	36	29	90
-24	$-\ 3$	-20	-14	-15	-20
35	60	77	22	14	70

50	48	66	74	85	96
-20	-37	-33	-10	-15	-23
30	11	33	64	70	73

80	34	55	70	25	61
-20	-21	-33	-40	-13	-21
60	13	22	30	12	40

Write the problems in columns. Subtract.

$56 - 2 =$	$75 - 20 =$	$50 - 30 =$
56	75	50
$-\ 2$	-20	-30
54	55	20

page 20 (twenty)

Page 21: Number Order, Number Symbols

CONCEPT(S): number order to 200, operation symbols

TEACHER GOAL(S): To teach the children
To review number order to 200, and
To review operation symbols (+, −, =, ≠, >, <).

MATERIALS/MANIPULATIVES:
pencils

TEACHING PAGE 21:
Turn to page 21 and read the directions to each exercise with the children. Remind them that the open side of the greater than and less than sign (>,<) is always toward the larger number. The children should be able to complete this page independently.

Write the number before and after.

42	43	44	197	198	199
15	16	17	75	76	77
119	120	121	49	50	51

Write the symbol (+, −) in the box.

15 $\boxed{-}$ 8 = 7   8 $\boxed{-}$ 5 = 3

6 $\boxed{+}$ 2 = 8   2 $\boxed{+}$ 9 = 11

Write the symbol (=, ≠) in the box.

3 − 3 $\boxed{\neq}$ 3   16 − 9 $\boxed{\neq}$ 8

7 + 5 $\boxed{=}$ 12   1 + 6 $\boxed{=}$ 7

Write the symbol (>, <) in the box.

129 $\boxed{<}$ 132   75 $\boxed{<}$ 85

115 $\boxed{<}$ 151   140 $\boxed{>}$ 139

Fill in the blanks.

<u>H</u> <u>A</u> <u>M</u> <u>B</u> <u>U</u> <u>R</u> <u>G</u> <u>E</u> <u>R</u>

U is the fifth letter.   G is the seventh letter.
A is the second letter.   H is the first letter.
E is the eighth letter.   R is the sixth letter.
M is the third letter.   B is the fourth letter.
R is the ninth letter.

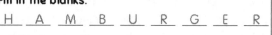

page 21 (twenty-one)

SELF TEST 3:

CONCEPT(S): directions, graphs, number symbols, writing vertical problems

TEACHER GOAL(S): To teach the children To learn to check their progress periodically.

MATERIALS/MANIPULATIVES:
pencils, crayons

TEACHING PAGE 22:

Turn to page 22. Read the directions with the children. Be sure they understand what they are to do. You may repeat the directions but give no other help. Do not have the children check their own work. Check it as soon as you can and go over it with each child. Show him where he did well and where he needs extra help.

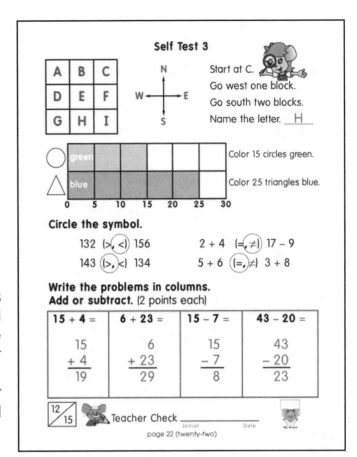

Self Test 3

A	B	C
D	E	F
G	H	I

N
W ——|—— E
S

Start at C.
Go west one block.
Go south two blocks.
Name the letter. __H__

Color 15 circles green.
Color 25 triangles blue.

Circle the symbol.

132 (>, <) 156   2 + 4 (=, ≠) 17 − 9

143 (>, <) 134   5 + 6 (=, ≠) 3 + 8

Write the problems in columns.
Add or subtract. (2 points each)

15 + 4 =	6 + 23 =	15 − 7 =	43 − 20 =
15	6	15	43
+ 4	+ 23	− 7	− 20
19	29	8	23

12/15   Teacher Check _____
Initial   Date
page 22 (twenty-two)

IV. PART FOUR

Page 23: Fractions, Number Place

CONCEPT(S): fractions; ones', tens', and hundreds' place

TEACHER GOAL(S): To teach the children
   To write fractions as part of a single object and part of a set, and
   To identify ones', tens', and hundreds' place in numbers.

MATERIALS/MANIPULATIVES:
pencils, shapes cut from construction paper - circle into four parts, rectangle into eight parts, triangle into two parts, objects for counting to represent four balloons, six cars, eight sodas

TEACHING PAGE 23:

   Place the shapes in front of the students and have them orally describe different parts of each shape. For example:
circle: *one part* is $\frac{1}{4}$ because the circle is divided into *four* parts and you are describing *one* part of the circle; *three parts* are $\frac{3}{4}$ because the circle is divided into *four* parts and you are describing *three* parts. Continue in this manner until the children can describe the parts correctly.

   Turn to page 23. Read the first direction and have the children complete that activity.

   Read the next directions. Place four objects in front of the children and say that these objects represent the set of four balloons. Have the children look at the first illustration. Ask how many parts the set of

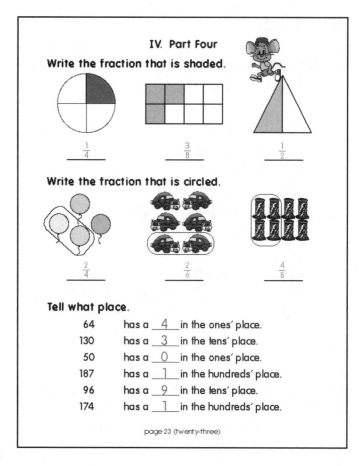

balloons is divided into. Tell them to show the same illustration with their objects. Say, "The set is divided into *4* balloons. (This is the denominator—bottom number of the fraction.) We are talking about *2* of the balloons. (This is the numerator—top number of the fraction.) The fraction is written $\frac{2}{4}$." Have the children complete this exercise using objects for counting.

   Children have difficulty understanding that fractions can be parts of a whole or parts of a set. Have them use the objects and shapes to illustrate a variety of fractions. Spend some time on this exercise.

   Read the final instructions. The students should be able to complete this exercise independently.

Page 24: Story Problems

CONCEPT(S): combining math skills and reading comprehension, counting money

TEACHER GOAL(S): To teach the children
 To combine math skills with reading comprehension, and
 To convert money to coins.

MATERIALS/MANIPULATIVES:
pencils, pennies, nickels, dimes

TEACHING PAGE 24:
 Turn to page 24 and read the title of the page with the children. Have them read the problems and then illustrate them with the pennies, nickels, and dimes. Tell the children to write the answers on the blanks. When the first two problems are complete, have the children look at the illustration. Tell them to write a story problem and then to write the problem using number symbols. They should complete the page by writing the numbers in order from smallest to largest.

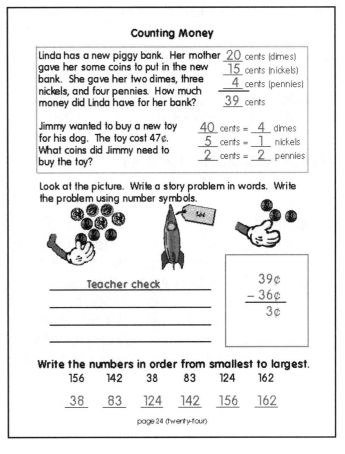

Page 25: Calendar

CONCEPT(S): reading a calendar

TEACHER GOAL(S): To teach the children
   To read a calendar.

MATERIALS/MANIPULATIVES:
pencils, current calendar

TEACHING PAGE 25:
   Turn to page 25 and ask the children to identify the illustration on the page. Direct their attention to the questions below the December calendar and have them write the answers. Allow them to complete this page as independently as possible. Discuss their answers when the page is complete.

December

Sunday	Monday	Tuesday	Wednesday	Thursday	Friday	Saturday
	1	2	3	4	5	6
7	8	9	10	11	12	13
14	15	16	17	18	19	20
21	22	23	24	25	26	27
28	29	30	31			

Name the month of the calendar.   December

Fill in the missing numbers on the calendar.

How many days are in this month?   31

What day is the first day of the month?   Monday

What day is the last day of the month?   Wednesday

Does everyone have snow in December?   no

What is the date today? **Teacher check**

_____   _____   _____   _____
(day of week)   (month)   (date)   (year)

page 25 (twenty-five)

Page 26: Telling Time

CONCEPT(S): time to five minutes, A.M., P.M.

TEACHER GOAL(S): To teach the children
To tell time to five minutes, and
To identify A.M. and P.M.

MATERIALS/MANIPULATIVES:
pencils, clock for student use

TEACHING PAGE 26:
Turn to page 26 and read the title with the children. Review clockwise motion with the students. Ask them to identify the hour hand and the minute hand on their clocks. Tell them to move the hands on their clocks in a clockwise motion. Have the children count by *5's* as they move the minute hand to each number on the clock from *1* to *12*. Write the time 6:40 on the board and ask the children to show the time on their clocks. Be sure the hour hand is between the 6 and 7. Remind the children that the hour is always the number that the hour hand is moving away from. Illustrate several other times to be sure the students understand. Have the children write the time for each clock on page 26 on the blank below the clock. Read the directions to the second activity. Have the children talk about A.M. and P.M. Tell them to give sensible answers to the questions.

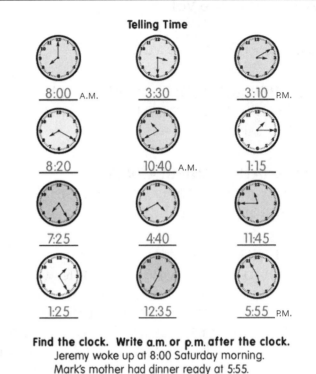

Telling Time

8:00 A.M.   3:30   3:10 P.M.

8:20   10:40 A.M.   1:15

7:25   4:40   11:45

1:25   12:35   5:55 P.M.

Find the clock. Write a.m. or p.m. after the clock.
Jeremy woke up at 8:00 Saturday morning.
Mark's mother had dinner ready at 5:55.
Lisa finished her spelling at 10:40.
David went to his friend's house at 3:10.

page 26 (twenty-six)

Page 27: Place Value

CONCEPT(S): place value to 199

TEACHER GOAL(S): To teach the children To write how many 1's, 10's, and 100's in a number.

MATERIALS/MANIPULATIVES:
pencils, objects for counting that represent 1's, 10's, 100's

TEACHING PAGE 27:
Turn to page 27. Read the title with the children and the instructions to the first exercise. Point to the 146 and ask the children what places the 1, the 4, and the 6 are in (hundreds', tens', ones'). Ask the children to write the value that the 1 represents (100), the 4 represents (40), the 6 represents (6). Students who understand the concept may go on to complete the page. Children who do not should use objects for counting to illustrate each number before writing the value on each blank. *Write the number* is completed in the same manner but in reverse order.

Place Value

Write the 100's, 10's, 1's.

		100's		10's		1,s
146	=	100	+	40	+	6
27	=		+	20	+	7
130	=	100	+	30	+	0
109	=	100	+	0	+	9
6	=		+		+	6
154	=	100	+	50	+	4

Write the number.

75	=		+	70	+	5
130	=	100	+	30	+	0
8	=		+		+	8
107	=	100	+	0	+	7
56	=		+	50	+	6
172	=	100	+	70	+	2

page 27 (twenty-seven)

Page 28: Sequence and Patterns

CONCEPT(S): identifying a pattern, naming what comes next in the sequence

TEACHER GOAL(S): To teach the children
To identify a pattern, and
To recognize and identify what comes next in the sequence.

MATERIALS/MANIPULATIVES:
pencils, crayons

TEACHING PAGE 28:
Turn to page 28 and read the directions at the top of the page with the children.

Allow the children to complete the page by drawing or writing what comes next in the sequence. This exercise develops the students' ability to think logically. Have them work as independently as possible.

Find the pattern. Write or draw what comes next.

150 ,	152,	154		156
7 $+ 2$ 9	2 $+ 7$ 9	9 $- 2$ 7		9 $- 7$ 2
12 $- 8$ 4	12 $- 7$ 5	12 $- 6$ 6		12 $- 5$ 7

page 28 (twenty-eight)

SELF TEST 4:

CONCEPT(S): coins, calendar, clocks, number expansion

TEACHER GOAL(S): To teach the children To learn to check their progress periodically.

MATERIALS/MANIPULATIVES: pencils

TEACHING PAGE 29:

Turn to page 29. Read the directions with the children. Be sure they understand what they are to do. You may repeat the directions but give no other help. Do not have the children check their own work. Check it as soon as you can and go over it with each child. Show him where he did well and where he needs extra help.

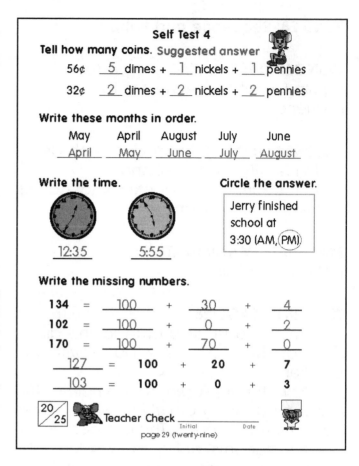

Self Test 4
Tell how many coins. Suggested answer

56¢ __5__ dimes + __1__ nickels + __1__ pennies

32¢ __2__ dimes + __2__ nickels + __2__ pennies

Write these months in order.

May	April	August	July	June
April	_May_	_June_	_July_	_August_

Write the time.   **Circle the answer.**

12:35   5:55

Jerry finished school at 3:30 (AM, (PM))

Write the missing numbers.

134 = __100__ + __30__ + __4__

102 = __100__ + __0__ + __2__

170 = __100__ + __70__ + __0__

__127__ = 100 + 20 + 7

__103__ = 100 + 0 + 3

20/25   Teacher Check _____ Initial ___ Date

page 29 (twenty-nine)

V. PART FIVE

Page 30: Measuring, Estimation

CONCEPT(S): measuring to the half inch, estimation

TEACHER GOAL(S): To teach the children
To measure lines to the half inch, and
To estimate the number of objects in a given area.

MATERIALS/MANIPULATIVES:
pencils, ruler to 12 inches, crayons, red construction paper, scissors, glue

TEACHING PAGE 30:

Turn to page 30 and read the rhyme with the children. Have them find the letters *A* to *M* on the picture. Read the instructions. Tell the children to use the ruler to draw lines connecting the letters as shown in the instructions. Have them measure the lines and write the inches on the blanks. Point to the apple in the upper right hand corner of the page. Ask the children how many apples they think the truck will hold and write the answer to the first statement. Children may draw apples in the truck using crayons, or they may cut out apples from construction paper using the illustrated apple as a pattern and glue them to the truck. When they have completed this part of the activity, ask them to count how many apples they actually put in the truck and to write the answer at the bottom of the page.

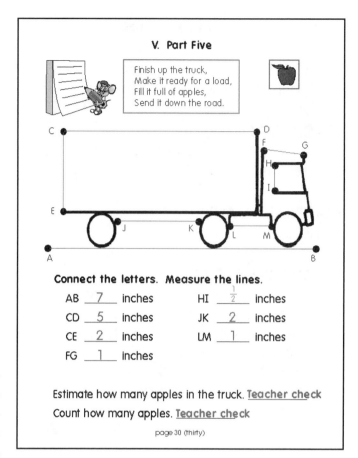

V. Part Five

Finish up the truck,
Make it ready for a load,
Fill it full of apples,
Send it down the road.

Connect the letters. Measure the lines.

AB __7__ inches   HI __$\frac{1}{2}$__ inches
CD __5__ inches   JK __2__ inches
CE __2__ inches   LM __1__ inches
FG __1__ inches

Estimate how many apples in the truck. Teacher check
Count how many apples. Teacher check

page 30 (thirty)

Page 31: Dictation Exercises

CONCEPT(S): applying listening and writing skills to math problems

TEACHER GOAL(S): To teach the children To write a mathematics problem from oral instruction.

MATERIALS/MANIPULATIVES: pencils

TEACHING PAGE 31:
Turn to page 31. This activity tests the students' ability to listen to a problem and transfer it on paper. The problems should be dictated with answers and in the manner shown so that the students use number symbols or number words.
Vertical:

```
  6      20      67      24
  3    + 30    + 31      31
+ 5      50      98    + 52
 14                     107


 13      50      46      87
- 5    - 30    - 3     - 22
  8      20      43      65
```

Horizontal:
Three plus eight equals eleven.
Fourteen minus nine equals five.

$$63 + 12 \neq 52 \qquad 56 - 31 = 25$$
$$18 > 12 \qquad 6 - 3 < 8 - 4$$

Fractions:
Circle:
Illustrate and write the fraction for $\frac{1}{4}$.
Rectangle:
Illustrate and write the fraction for $\frac{4}{6}$.
Triangle:
Illustrate and write the fraction for $\frac{1}{2}$.

Listen and write.

6 3 + 5 — 14	20 + 30 — 50	67 + 31 — 98	24 31 + 52 — 107
13 - 5 — 8	50 - 30 — 20	46 - 3 — 43	87 - 22 — 65

Three plus eight equals eleven.	
Fourteen minus nine equals five.	
$63 + 12 \neq 52$	$56 - 31 = 25$
$18 > 12$	$6 - 3 < 8 - 4$

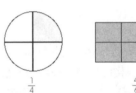

$$\frac{1}{4} \qquad \frac{4}{6}$$

$$\frac{1}{2}$$

page 31 (thirty-one)

Page 32: Fractions

CONCEPT(S): writing parts of sets as fractions

TEACHER GOAL(S): To teach the children
To write parts of sets as fractions, and
To illustrate a fraction as part of a single object.

MATERIALS/MANIPULATIVES:
pencils, objects for counting

TEACHING PAGE 32:

Turn to page 32. Read the instructions. Have the children count the number of objects in the first set. Ask if this is the same as the bottom number of the fraction. Ask the children how many of the objects they should circle to show $\frac{1}{4}$. Allow the children to complete the page but monitor their work carefully so they do not develop a pattern of error.

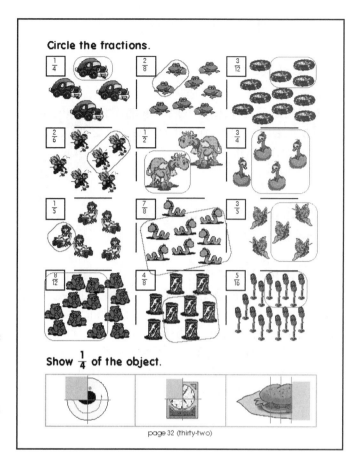

page 32 (thirty-two)

Page 33: Number Order

CONCEPT(S): number order to 200

TEACHER GOAL(S): To teach the children
To identify number order to 200, and
To recognize sequence in number facts.

MATERIALS/MANIPULATIVES:
pencils

TEACHING PAGE 33:
Turn to page 33. Read the directions for each exercise with the children. The exercises on this page contain number order skills that the students should have mastered. Allow them to complete the page independently.

Write the numbers in order from smallest to largest.

143	127	38	172	163	138
38	127	138	143	163	172

Write the numbers in order from largest to smallest.

3	9	6	7	0	2	5	1	8	4
9	8	7	6	5	4	3	2	1	0

Write the missing numbers.
Count by 1's, 2's, 5's, or 10's.

123 _124_ _125_ 126   12 _14_ _16_ 18
55 _60_ _65_ 70   20 _30_ _40_ 50

Write the numbers before and after.

_78_ 79 _80_   _126_ 127 _128_
_10_ 11 _12_   _188_ 189 _190_

Write the next fact.

$$\begin{array}{ccc} 6 & 6 & 6 \\ +3 & +4 & +5 \\ \hline 9 & 10 & 11 \end{array} \quad \begin{array}{ccc} 11 & 11 & 11 \\ -7 & -6 & -5 \\ \hline 4 & 5 & 6 \end{array} \quad \begin{array}{ccc} 8 & 8 & 8 \\ +1 & +2 & +3 \\ \hline 9 & 10 & 11 \end{array}$$

page 33 (thirty-three)

SELF TEST 5:

CONCEPT(S): measurements, oral response, identifying fractions

TEACHER GOAL(S): To teach the children To learn to check their progress periodically.

MATERIALS/MANIPULATIVES:
pencils, rulers

TEACHING PAGE 34:

Turn to page 34. Read the directions with the children. Be sure they understand what they are to do. You may repeat the directions but give no other help. Do not have the children check their own work. Check it as soon as you can and go over it with each child. Show him where he did well and where he needs extra help.

Listen and write: The problems should be dictated with answers and in the manner shown so that the students use number symbols or number words.
Vertical:

```
  4      78      87      90
  2    + 31    - 33    - 40
+ 7     109      54      50
 13
```

Horizontal:

Nine plus four equals thirteen.

$43 + 22 \neq 75$ $56 > 31$

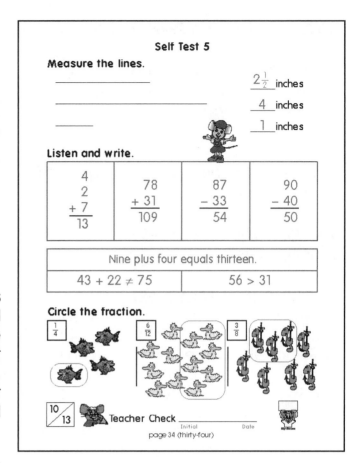

LIFEPAC TEST AND ALTERNATE TEST 110

CONCEPT(S): addition, subtraction, place value, time, number words, fractions, listening skills

TEACHER GOALS: To teach the children To learn to check their own progress periodically.

MATERIALS/MANIPULATIVES:
pencils, rulers

TEACHING the LIFEPAC TEST:

Administer the test in at least two sessions.

Read all of the directions on each page as the children prepare to do it. Be sure that they understand what they are being asked to do. Give no help except with directions. Go over each page with the child as soon as possible after you check it so that he can see where he did well and where he needs more work.

Evaluate the tests and review areas where the children have done poorly. Review the pages and activities that stress the concepts tested.

If necessary, when the children have reviewed sufficiently, administer the Alternate LIFEPAC test. Follow the same procedures as used for the LIFEPAC Test.

LIFEPAC Test
Listen and write. (page 2)

Problems should be dictated with answers and in the manner shown so that the students use number symbols or number words.
Dictate:

22	30	15	46
5	51	− 7	− 31
+ 30	+ 7	8	15
57	88		

Five plus nine equals fourteen.

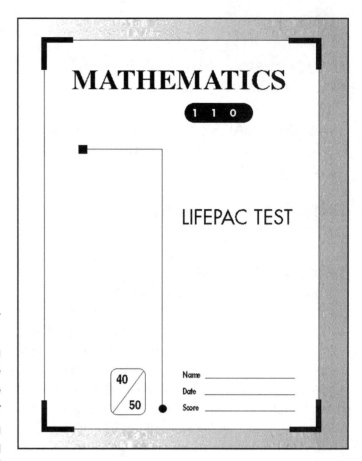

MATHEMATICS 110: LIFEPAC TEST

Write a family of facts for 3, 7, 10. (4 points)

+ 3 7	+ 7 3	− 10 3	− 10 7
10	10	7	3

Even numbers end in __0__, __2__, __4__, __6__, and __8__.

Write the answer.

Is 7 closer to 0 or 10? __10__   Is 22 closer to 20 or 30? __20__

Seven plus four equals ____eleven____.
Fourteen minus eight equals ____six____.

Circle the correct answer.

seventy-four (>̲ , <) sixty-two   4 + 8 (=̲ , ≠) 6 + 6
3 + 6 (>̲ , <) 17 − 9   three plus five (= , ≠̲) nine

1. Draw a set of four squares.
2. Circle $\frac{3}{4}$ of the set. (2 points)

page 1 (one)

Alternate LIFEPAC Test
Listen and write. (page 2)
Problems should be dictated with answers and in the manner shown so that the students use number symbols or number words.
Dictate:

```
  40      71      12      69
  12      20     − 8     − 53
 + 4     + 7       4      16
  56      98
```

Seventeen minus eight equals nine.

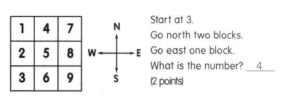

Start at 3.
Go north two blocks.
Go east one block.
What is the number? 4
(2 points)

Tell how many coins. (each answer, 1 point) Suggested answers.

76¢ _7_ dimes + _1_ nickels + _1_ pennies
43¢ _4_ dimes + _0_ nickels + _3_ pennies

Write the time. (each answer, 1 point) **Circle the answer.**

8:20   10:05

Sara ate breakfast at 8:00 (A.M.), P.M.)

Listen and write. (each answer, 1 point)

22 5 +30 57	30 51 + 7 88	15 − 7 8	46 −31 15
Five plus nine equals fourteen.			

page 2 (two)

Write the missing numbers. (8 points)

156 = _100_ + _50_ + _6_
107 = _100_ + _0_ + _7_
_135_ = **100** + **30** + **5**
_72_ = + **70** + **2**

Add or subtract.

```
  3    23    42    16    15    49    58
  5    32     3   − 7   − 9   − 6   −23
 + 8  +41   +22    9     6    43    35
 16    96    67
```

page 3 (three)

215

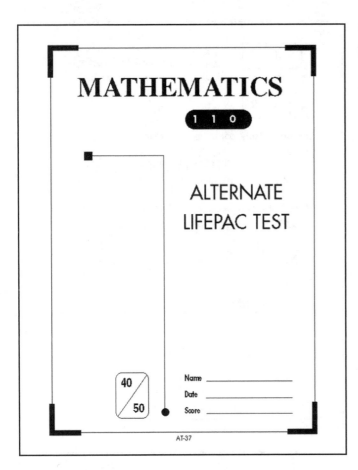

MATHEMATICS
1 1 0

ALTERNATE
LIFEPAC TEST

40 / 50

Name _____
Date _____
Score _____

AT-37

MATHEMATICS 110: Alternate LIFEPAC TEST

Write a family of facts for 5, 6, 11. (4 points)

+ 5	+ 6	− 11	− 11
6	5	6	5
11	11	5	6

Even numbers end in __0__, __2__, __4__, __6__, and __8__.

Write the answer.

Is 4 closer to 0 or 10? __0__   Is 27 closer to 20 or 30? __30__

Eight plus six equals ___fourteen___.
Twelve minus five equals ___seven___.

Circle the correct answer.

fifty-eight (>) , <) thirty-nine   3 + 9 (=) , ≠) 8 + 2

4 + 9 (>) , <) 5 + 6   two plus nine (=) , ≠) eleven

1. Draw a set of three triangles. 2. Circle $\frac{2}{3}$ of the set. (2 points)	△ △ △

AT-38

1	4	7
2	5	8
3	6	9

N
W — E
S

Start at 7
Go west two blocks.
Go south two blocks.
What is the number? __3__
(2 points)

Tell how many coins. (6 points) Suggested answers.

64¢ __6__ dimes + __0__ nickels + __4__ pennies
89¢ __8__ dimes + __1__ nickels + __4__ pennies

Write the time.   **Circle the answer.**

Sara ate dinner at 6:00 (A.M., (P.M.)

9:35   4:05

Listen and write.

40 12 + 4 56	71 20 + 7 98	12 − 8 4	69 −53 16
Seventeen minus eight equals nine.			

237

Write the missing numbers. (8 points)

35	=	_____	+	__30__	+	__5__
129	=	__100__	+	__20__	+	__9__
__180__	=	100	+	80	+	0
__34__	=		+	30	+	4

Add or subtract.

7 2 + 9 18	43 21 + 35 99	62 5 + 40 107	15 − 8 7	13 − 5 8	56 − 4 52	89 − 46 43

AT-40

T
E
S
T
S

Reproducible Tests
for use with the Mathematics
100 Teacher's Guide

MATHEMATICS

106

ALTERNATE
LIFEPAC TEST

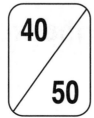

Name _____

Date _____

Score _____

MATHEMATICS 106: Alternate LIFEPAC TEST

Write the answer to the facts.

$$\begin{array}{r} 6 \\ + 5 \\ \hline \end{array} \qquad \begin{array}{r} 9 \\ + 9 \\ \hline \end{array} \qquad \begin{array}{r} 4 \\ + 8 \\ \hline \end{array} \qquad \begin{array}{r} 10 \\ - 3 \\ \hline \end{array} \qquad \begin{array}{r} 8 \\ - 0 \\ \hline \end{array} \qquad \begin{array}{r} 7 \\ - 5 \\ \hline \end{array}$$

Count by 2's, 5's, 10's. (Each row 1 point)

2, ___, 6, ___, ___, 12, 14,

___, 10, ___, ___, 25, 30, ___

10, 20, ___, ___, ___, 60, 70

Circle the odd numbers. (2 points)

10   27   39   82   46   73   55

Find the sum. Add down. Add up.

$$\begin{array}{r} 3 \\ 0 \\ + 5 \\ \hline \end{array} \qquad \begin{array}{r} 3 \\ 6 \\ + 2 \\ \hline \end{array} \qquad \begin{array}{r} 5 \\ 3 \\ + 4 \\ \hline \end{array} \qquad \begin{array}{r} 6 \\ 2 \\ + 3 \\ \hline \end{array}$$

Write the symbol.

forty-five ____

thirty-three ____

Write the words.

62 _____

28 _____

Circle (=, ≠).

3 + 5 (=, ≠) 4 + 4

8 − 6 (=, ≠) 10 + 8

Circle (>, <).

12 (>, <) 13

54 (>, <) 43

Write a number sentence.

3, 8, 11 _____

seven, four three _____

Write how many.

70 = ___ tens + ___ ones   **3** tens + **6** ones = ___

Read. Write the problem.(1 pt.) **Label the answer.**(1 pt.)

Jim read 6 pages in his book on Monday, 3 pages on Tuesday, and 4 pages on Thursday. How many pages did he read altogether?

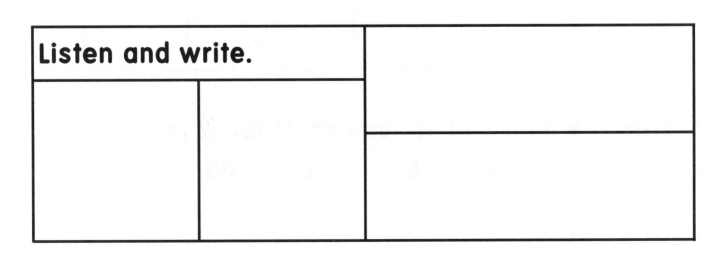

Listen and write.

Write the time on the clock.

_____   _____   _____

Write how many you need.

77¢   ____ dimes   ____ nickels   ____ pennies

29¢   ____ dimes   ____ nickels   ____ pennies

Measure the line.

_____   ____ inches

Write a fact to make the answer true.

_____	_____	_____	_____
8	2	6	13

Write the numbers in number order. (2 pts.)

82   12   65   4   43   29

____ ____ ____ ____ ____ ____

MATHEMATICS

107

ALTERNATE
LIFEPAC TEST

$\dfrac{40}{50}$

Name _____

Date _____

Score _____

MATHEMATICS 107: Alternate LIFEPAC TEST

Put a circle around each number greater than 148.

122   172   109   150   132   198

Put a circle around each number less than 159.

116   199   180   152   175   130

Write 100's, 10's, 1's. (6 pts.)

130 = _____ + _____ + _____

46 = _____ + _____ + _____

Write the number or number word.

thirteen _____   79 _____

eighty-six _____   60 _____

Write the answer on the line ___.

Mark spent 89¢ at the store.

He used ____ **dimes** ____ **nickels** ____ **pennies**

Write the answer.

$$\begin{array}{r} 7 \\ + 3 \\ \hline \end{array} \qquad \begin{array}{r} 4 \\ 3 \\ + 6 \\ \hline \end{array} \qquad \begin{array}{r} 7 \\ 1 \\ + 8 \\ \hline \end{array} \qquad \begin{array}{r} 32 \\ + 6 \\ \hline \end{array} \qquad \begin{array}{r} 42 \\ + 36 \\ \hline \end{array}$$

Write the answer.

$$\begin{array}{r} 10 \\ -\ 5 \\ \hline \end{array} \qquad \begin{array}{r} 12 \\ -\ 4 \\ \hline \end{array} \qquad \begin{array}{r} 11 \\ -\ 3 \\ \hline \end{array} \qquad \begin{array}{r} 11 \\ -\ 6 \\ \hline \end{array} \qquad \begin{array}{r} 12 \\ -\ 7 \\ \hline \end{array}$$

Circle =, ≠.

5 + 9 (=, ≠) 15   Seven plus nine (=, ≠) seventeen.

14 − 6 (=, ≠) 8   Six minus zero   (=, ≠) six.

Circle >, <.

163 (>, <) 158   6 + 2 (>, <) 5 + 4

194 (>, <) 200   9 − 4 (>, <) 11 − 9

Write the time.

_____

Write A.M. or P.M.

We eat
breakfast.

_____

Write the fraction.

 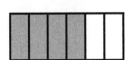 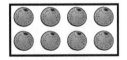 = _____ oranges

  = _____ dolls

_____   _____

225

Write a Big number and a Little number.

	Big	Little
7, 6	_____	_____
3, 1	_____	_____

Write how many. ①_____ ②_____ ③_____

226

MATHEMATICS

1 0 8

ALTERNATE LIFEPAC TEST

40 / 50

Name _____

Date _____

Score _____

MATHEMATICS 108: Alternate LIFEPAC TEST

Add down. Add up. (5 points)

$$\begin{array}{r} 4 \\ 2 \\ +6 \\ \hline \end{array} \qquad \begin{array}{r} 5 \\ 1 \\ +8 \\ \hline \end{array} \qquad \begin{array}{r} 46 \\ +2 \\ \hline \end{array} \qquad \begin{array}{r} 41 \\ +25 \\ \hline \end{array} \qquad \begin{array}{r} 63 \\ +24 \\ \hline \end{array}$$

Subtract.

$$\begin{array}{r} 18 \\ -9 \\ \hline \end{array} \qquad \begin{array}{r} 16 \\ -7 \\ \hline \end{array} \qquad \begin{array}{r} 14 \\ -8 \\ \hline \end{array} \qquad \begin{array}{r} 17 \\ -9 \\ \hline \end{array} \qquad \begin{array}{r} 16 \\ -7 \\ \hline \end{array}$$

Write 100's, 10's, 1's. (each row 1 point)

	100's	10's	1's
150 =	_____	+ _____	+ _____
65 =	_____	+ _____	+ _____
29 =	_____	+ _____	+ _____

Write the number.

_____ =   + 40 + 3

_____ = 100 + 70 + 5

Write the time.

____ : ____   ____ : ____   ____ : ____   ____ : ____   ____ : ____

Write the number or number word.

35 _____   forty-one   _____

13 _____   sixty-three _____

152 _____   fifty   _____

Measure.

_____   ____ inches

___   ____ inches

Add or subtract.
Find the answer on the number line. Write the numbers from the number line that are before and after. Circle the closest number. (each problem 3 pts.)

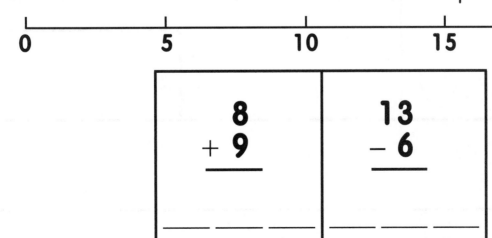

229

Write the fraction.

_____   _____

_____   _____

Solve the problem. (2 pts)

Tony had eight nickels in his pocket. His sister borrowed five. How many nickels does Tony have now?

_____ nickels

Listen and write. (5 points.)

Listen and write. (5 points.)

MATHEMATICS

1 0 9

ALTERNATE

LIFEPAC TEST

40 / 50

Name _____

Date _____

Score _____

MATHEMATICS 109: LIFEPAC TEST

Complete the fact families for addition and subtraction. (2 points each)

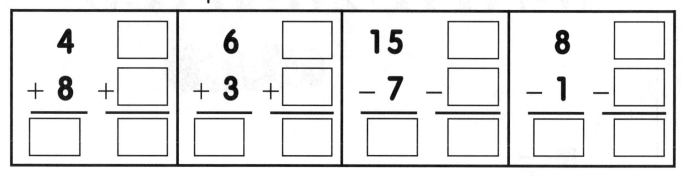

$$\begin{array}{r} 4 \\ + 8 \\ \hline \end{array} \quad \begin{array}{r} \square \\ + \square \\ \hline \square \end{array} \qquad \begin{array}{r} 6 \\ + 3 \\ \hline \end{array} \quad \begin{array}{r} \square \\ + \square \\ \hline \square \end{array} \qquad \begin{array}{r} 15 \\ - 7 \\ \hline \end{array} \quad \begin{array}{r} \square \\ - \square \\ \hline \square \end{array} \qquad \begin{array}{r} 8 \\ - 1 \\ \hline \end{array} \quad \begin{array}{r} \square \\ - \square \\ \hline \square \end{array}$$

Add or subtract.

$$\begin{array}{r} 3 \\ 2 \\ + 5 \\ \hline \end{array} \qquad \begin{array}{r} 73 \\ + 6 \\ \hline \end{array} \qquad \begin{array}{r} 45 \\ 30 \\ + 21 \\ \hline \end{array} \qquad \begin{array}{r} 13 \\ - 5 \\ \hline \end{array} \qquad \begin{array}{r} 67 \\ - 4 \\ \hline \end{array} \qquad \begin{array}{r} 83 \\ - 71 \\ \hline \end{array}$$

Write the number word.

12 _____   22 _____

Write the days in order. (3 points)

Saturday   Monday   Tuesday   Friday   Thursday   Wednesday   Sunday

_____   _____   _____   _____   _____   _____   _____

Write the time.

___ : ___   ___ : ___   ___ : ___   ___ : ___

Write the answer.

Is 8 closer to 0 or 10? ＿＿＿＿   Is 3 closer to 0 or 10? ＿＿＿＿

Tammie has five dimes, two nickels, and six pennies in her hand. How much money does she have? (2 points)

＿＿＿＿＿ cents (dimes)

＿＿＿＿＿ cents (nickels)

＿＿＿＿＿ cents (pennies)

＿＿＿＿＿ cents

Circle =, ≠, >, <.

$9 + 3$ (=, ≠) 11 $8 - 5$ (>, <) $4 + 0$

$14 - 7$ (=, ≠) $3 + 4$ $12 - 7$ (>, <) $16 - 8$

Listen and write. (2 points each)

Circle the correct answer.

Jenny spent the day at her friends house. She left home at 9:00 P.M. and came home at 3:00 A.M.

(sensible, nonsense)

Write the fractions. Tell the shape.

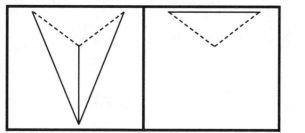

_____ _____

_____

Write the fractions. Name the set.

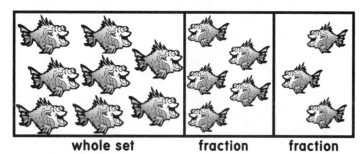

whole set   fraction   fraction

_____ _____

_____

Write the answer.

39 has a _____ in the tens' place.

106 has a _____ in the ones' place.

170 has a _____ in the hundreds' place.

5 has a _____ in the ones' place.

MATHEMATICS

110

ALTERNATE LIFEPAC TEST

40 / 50

Name _____

Date _____

Score _____

MATHEMATICS 110: Alternate LIFEPAC TEST

Write a family of facts for 5, 6, 11. (4 points)

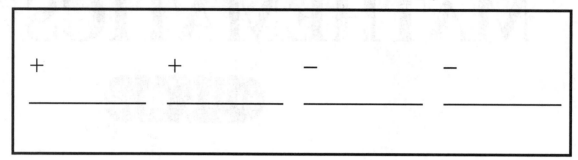

+	+	−	−
_____	_____	_____	_____

Even numbers end in ____, ____, ____, ____, and ____.

Write the answer.

Is 4 closer to 0 or 10? ____   Is 27 closer to 20 or 30? ____

Eight plus six equals _____.

Twelve minus five equals _____.

Circle the correct answer.

fifty-eight (>, <) thirty-nine   3 + 9 (=, ≠) 8 + 2

4 + 9 (>, <) 5 + 6   two plus nine (=, ≠) eleven

1. **Draw a set of three triangles.**
2. **Circle $\frac{2}{3}$ of the set.**
   (2 points)

1	4	7
2	5	8
3	6	9

N
W ← → E
S

Start at 7.

Go west two blocks.

Go south two blocks.

What is the number? _____

(2 points)

Tell how many coins. (6 points)

64¢   ____ dimes + ____ nickels + ____ pennies

89¢   ____ dimes + ____ nickels + ____ pennies

Write the time.

___ : ___   ___ : ___

Circle the answer.

Sara ate dinner at

6:00 (A.M., P.M.)

Listen and write.

Write the missing numbers. (8 points)

$35 = \underline{\hspace{3cm}} + \underline{\hspace{3cm}} + \underline{\hspace{2cm}}$

$129 = \underline{\hspace{3cm}} + \underline{\hspace{2cm}} + \underline{\hspace{1.5cm}}$

$\underline{\hspace{3cm}} = 100 + 80 + 0$

$\underline{\hspace{3cm}} = + 30 + 4$

Add or subtract.

```
   7      43      62      15      13      56      89
   2      21       5    - 8     - 5     - 4    - 46
 + 9    + 35    + 40    ____    ____    ____    ____
```

238